Agriculture in
Ante-Bellum
Mississippi

SOUTHERN CLASSICS SERIES

Mark M. Smith and Peggy G. Hargis, Series Editors

Agriculture in *Ante-Bellum* Mississippi

John Hebron Moore

New Introduction by Douglas Helms

The University of South Carolina Press

Published in Cooperation with the Institute for
Southern Studies of the University of South Carolina

Cloth edition published by Bookman Associates, New York
Paperback edition published by the University of South Carolina Press,
Columbia, South Carolina 29208

www.sc.edu/uscpress

Manufactured in the United States of America

19 18 17 16 15 14 13 12 11 10   10 9 8 7 6 5 4 3 2 1

Library of Congress Cataloging-in-Publication Data

Moore, John Hebron.
    Agriculture in ante-bellum Mississippi / John Hebron Moore ; new
introduction by Douglas Helms.
        p. cm. — (Southern classics series)
    "Published in cooperation with the Institute for Southern Studies of the
University of South Carolina."
    Originally published: New York : Bookman Associates, 1958.
    Includes bibliographical references and index.
    ISBN 978-1-57003-877-8 (pbk. : alk. paper)
    1. Agriculture—Mississippi—History—18th century. 2. Agriculture—
Mississippi—History—19th century. 3. Agricultural innovations—
Mississippi—History—18th century. 4. Agricultural innovations—
Mississippi—History—19th century. 5. Cotton growing—Mississippi—
History—18th century. 6. Cotton growing—Mississippi—History—
19th century. 7. Mississippi—Economic conditions—18th century.
8. Mississippi—Economic conditions—19th century. I. University of
South Carolina. Institute for Southern Studies. II. Title.
    S451.M7M6 2010
    630.9762'09034—dc22

                                                                    2009022717

This book was printed on Glatfelter Natures, a recycled paper with 30
percent postconsumer waste content.

Publication of the Southern Classics series is made possible in part by the
generous support of the Watson-Brown Foundation.

Contents

Series Editors' Preface

John Hebron Moore's classic study, *Agriculture in Ante-Bellum Missis-sippi*, reminds us of the braided, interconnected nature of antebellum southern agriculture. Southern cotton was blown by winds economic, political, and cultural; what happened in the nation's capital, in the economy generally, and in the realm of technology mattered to the Mississippi planter. Moore's planters were thoughtful men, with an eye to profit and a willingness to adopt new methods and technologies to help achieve it.

Originally published in 1958, *Agriculture in Ante-Bellum Mississippi* has become a staple in the historiography of the antebellum South, and anyone looking for a first-rate economic history of cotton cultivation—detailing how it was grown, how it was ginned, and why it was so important—in one of the Old South's most important cotton-producing states is well advised to read the study with care. Douglas Helm's insightful new introduction helps us understand a good deal about Moore the teacher, the historian, and the continuing relevance of this classic study.

Southern Classics returns to general circulation books of importance dealing with the history and culture of the American South. Sponsored by the Institute for Southern Studies, the series is advised by a board of distinguished scholars who suggest titles and editors of individual volumes to the series editors and help establish priorities in publication.

Chronological age alone does not determine a title's designation as a Southern Classic. The criteria also include significance in contributing to a broad understanding of the region, timeliness in relation to events and moments of peculiar interest to the American South, usefulness in the classroom, and suitability for inclusion in personal and institutional collections on the region.

MARK M. SMITH
PEGGY G. HARGIS
Series Editors

Introduction

In his classes on the Old South at Florida State University, John Hebron Moore approached the lectern armed only with a note card, if that much. He made no show of it, but we noticed. The lecture or lectures that impressed and stuck in the memory above all others featured the development of the cotton culture in the Natchez District. The story did not fit the model. This was no tale of the gradual settlement westward by yeoman cotton farmers from the Atlantic coast. Rather, during the 1790s settlers of means intent on raising cotton skipped over the rest of the South and moved directly to the loess bluffs overlooking the Mississippi River near Natchez. The deep, wind-deposited soil had supported the Natchez tribe before Europeans arrived, and England had begun encouraging settlement after 1763 when it acquired West Florida at the conclusion of the Seven Years' War. Then a confluence of several factors touched off the mania for cotton around 1795. Rebellions in the Caribbean disrupted supply just as mechanization of the textile industry in England was creating greater demand. In addition to its fertile soils, Natchez was blessed with river transport and the long growing season required by the cotton plant. Moore's epic account continued with the farmers acquiring a drawing of the Whitney cotton gin. A planter's son and a slave fashioned circular saw-toothed blades and substituted them for the wire rods of the Whitney design. This modified Whitney gin became the standard for southern saw gins. Planters experimented with cottons suited to the loess lands and eventually crossed the Green Seed cotton with cotton acquired from the plateau of central Mexico. The so-called Mexican cotton, especially the selection known as Petit Gulf, became the standard upland cotton of the South.

The cotton kingdom of the Old Southwest expanded until the issuance of Andrew Jackson's "specie circular," which triggered the depression of 1837–49. The price of cotton peaked at twenty-one cents a pound in 1836 and then slid to its nadir, five cents a pound, in 1845. Planters responded by cutting costs, in effect becoming more efficient. They produced much of their feed and food, and they reduced purchases from the Midwest.

They learned the magic of fixing nitrogen in the soil by planting legumes and adopted soil conservation measures. Due to their effective response to the depression, they entered the 1850s strengthened and were able to produce more cotton and produce it more cheaply than they had before the depression. The concentration on cotton growing intensified, and the acquisition of slaves in Mississippi quickened. During the 1850s the slave population increased 41 percent while the white population increased only 20 percent. The trend showed no sign of abating on the eve of the Civil War.

This fascinating tale could be found in *Agriculture in Ante-Bellum Mississippi*, originally issued by Bookman Associates in 1958, reissued by Octagon Books in 1971, and now reprinted by the University of South Carolina Press.[1] The growing and ginning of cotton played a central role in the book, and Moore had grown up in a cotton culture. Yet he had taken a most circuitous route to studying the agricultural history of the state. John Hebron Moore was from the Mississippi Delta, born in Greenville on February 26, 1920. His childhood would be unsettled, but not one of want. His maternal grandfather, John Lawrence Hebron, was a prominent planter, lawyer, and politician in Leland, Mississippi, who had campaigned for the abolition of the convict lease system as inhumane and for the establishment of a state penal institution. Hebron furthermore secured his son-in-law a job as bookkeeper and cotton classifier at the new Parchman penitentiary farm. Young John Hebron Moore lived on the Parchman farm until his mother died when he was about nine years old. His father and grandparents then decided that it was best for him to live with his grandparents in Leland. Moore's teenage years would be spent between Leland and Jackson, where his grandfather was in the state legislature. As his grandmother's condition demanded more medical attention, it was decided he and she should live near the hospital at Jackson.

At Mississippi State College (now Mississippi State University), Moore had a reserve commission in the army. By September 1941, when the army inducted him into the quartermaster corps, he had completed his junior year in the aeronautical engineering curriculum. After the war he completed his degree at Mississippi State and embarked on a career in the aircraft industry. Understandably the industry grew with wartime demand; then came the crash. Moore worked first for the North American Aviation Company in Los Angeles. As the government contracts expired, layoffs loomed and Moore took a similar job with the Curtiss-Wright Corporation in Columbus, Ohio. That company similarly lost much of its business, and

Moore returned to Mississippi where he worked temporarily in a relative's ice business while he contemplated his future. He decided to try to make a living doing what he did for recreation, which was to read history.

After earning a masters degree at the University of Mississippi (1951), he enrolled in the graduate history program at Emory University. The inspiration to study antebellum agriculture came in an Old South seminar taught by James C. Bonner, a visiting professor from Georgia State College for Women at Milledgeville. Bonner was studying antebellum agricultural history and would publish *A History of Georgia Agriculture, 1732–1860.* James Z. Rabun directed Moore's dissertation, "Mississippi Agriculture, 1770–1860," for which Moore was awarded a history doctorate in 1955. His was among the first doctorates awarded by the department. At Emory, John Moore met Margaret DesChamps of Bishopville, South Carolina, whom he would marry on December 21, 1955. She had earned the first history doctorate awarded by Emory University in 1952 for a dissertation titled "The Presbyterian Church in the South Atlantic States, 1801–1861."

Following a year of teaching at Delta State College in Cleveland, Mississippi (1955–56), he moved to the University of Mississippi in 1956 and was there when Bookman Associates of New York City published *Agriculture in Ante-Bellum Mississippi* in 1958.

Agriculture in Ante-Bellum Mississippi took the entire state as its subject area, not just the Natchez District. Settlement of North Mississippi began in earnest in 1833, more than three decades after the cotton mania swept the Natchez District. After lands ceded by the Choctaws and Chickasaws in the treaties of Pontotoc and Dancing Rabbit Creek were placed on the market in 1833, land speculation, fueled by easy terms, triggered a flood of settlers from the Old South and south Mississippi to the "New Purchase." And the topics examined transcended cotton even though Moore asserted, "The economic history of Mississippi, more than that of any other state, is the history of cotton" (10). Livestock, grains, grasses, and food crops grew in importance as the farmers responded to the depression. Not every planter was interested in improved breeds, and Moore had to concede that "most planters were more interested in fine race horses, saddle horses, and carriage horses than they were in superior breeds of cattle, sheep, and hogs" (84). Nonetheless a small but talented and energetic group sought to improve the state's livestock and food and feed crops through purchase, selection, and breeding. These efforts at breeding might not have proved as financially remunerative as the selection of Petit Gulf cotton, but they too are part of agricultural reform in Mississippi.

Moore introduced readers to a remarkable group of agricultural innovators, men such as William Dunbar, Rush Nutt, Henry W. Vick, Thomas Affleck, Martin W. Philips, N. G. North, and his own ancestor John Hebron. Affleck, for example, campaigned for permanent Bermuda grass pastures, improved plant and animal breeds, and improved agricultural equipment. He also proposed crop diversification as a rational economic strategy. Others argued for maximum cotton production, even in the face of declining prices, so as to maintain a firm hold on foreign markets. John Hebron brought peaches and his slaves from Virginia to Mississippi, where he grew peaches for his family and the slaves. He sold ten barrels at five dollars a barrel to a free African American trader who sold them in New Orleans for twenty-nine dollars a barrel. The incident awakened John Hebron to "a better business than raising cotton." At the dawn of the Civil War, the "pear King" had a seventy-thousand-tree nursery and one hundred thousand mature pear trees from which he shipped pears to New Orleans, St. Louis, and Chicago (142).

Moore examined the state's indigenous agricultural implement industry, which manufactured tools and equipment for the cultivation of cotton and other crops confined to the South. Planters invested in railroads and agricultural implement manufacturing to consolidate the gains of the cotton kingdom. This volume made the case, as did his lectures, for greater appreciation of Old South innovation in the development of agricultural technologies. Had chairmanship of the history department and a heart attack in 1992 not intervened, it had been his goal to contribute more to the history of technology in the Old South.

Moore discussed the role of the agricultural press in educating farmers about new methods and technology. Agricultural improvement societies also sprang up during the reform movement. A corollary to the strengthening of the cotton economy was that some agricultural reform societies, born as a response to the depression, became centers for secession sentiment. Not all participants in the societies endorsed the secession movement.

In trying to assess the place of *Agriculture in Ante-Bellum Mississippi* in the historiography of the South, it is well to recall that Moore thought of his work as falling within economic history and that he called for more such studies of the state's transportation, finance, and manufacturing, and agricultural sectors. In his opinion most twentieth-century historians, obsessed with the causes of the Civil War, had neglected the study of Mississippi's economic and cultural history as they focused on the institution of slavery and on the state's political history. (He exempted the work of the antebellum Mississippi historians and writers Benjamin L. C. Wailes, John F. H. Claiborne, and Horace S. Fulkerson from this charge.)

Moore followed a chronological and topical approach in his study. Readers, especially students of the Old South, may differ as to his most significant contributions. The fact that there is room for choice highlights two attributes of the study. First, it speaks to the depth and breadth of his historical narrative that it allows for such selection. Second, Moore proposed no theory or novel interpretations; he did not set the terms on which to judge the study. For this reader his three most significance contributions examine the independent development of the Natchez District as a cotton kingdom apart from the cotton culture on the Atlantic coast, the evolution of Petit Gulf Cotton and antebellum cotton breeding, and the impetus the 1837–49 depression had on an agricultural reform movement that solidified the position of cotton as Mississippi's most profitable crop in the 1850s.

Moore's discussion of the development of cotton culture of the Natchez District surpassed all that had come before it. Agricultural history in the United States is not devoid of studies on the role of plant breeding. Still, both nationally and regionally for the South, labor-saving mechanical innovations have received more attention than the productivity gains of the biological revolution. Recently Moore's work on cotton breeding has received some attention in economic history.

Moore made a major contribution to southern history with his discussion of the agricultural depression of 1837–49 and its ramifications. Up to the issuance of the specie circular, the government had required payment of one-fourth of the price of public land at purchase, to be followed by three annual payments. The specie circular demanded full payment in specie, and that brought the financial structure down. It was not so much the distress caused by the depression as the planters' and farmers' reaction to the economic depression that Moore illuminated. That was the real significance in his view. Planters became more efficient and increased their profit margin. Part of the efficiency involved diversification and growing more feed and food at home. The alterations in cotton growing—soil conservation practices, greater use of labor saving machinery, plant selection—in response to the depression required more attention from the slaves. They became more skilled, and this was also part of the agricultural reform story. As cotton growing became more intensive and less extensive, the price of slaves—reflecting their productiveness in this new mode of operating—increased just as opposition to slavery was growing in the North.

Moore's childhood, spent immersed in a cotton culture, was no doubt an advantage in writing about the subject. Sometimes sentiments about hearth and home can trip one up in trying to write objectively, if that is in fact possible. His rather stark assessments prove his upbringing proved no obstacle. From a native Mississippian, who inherited family land and who still farms it, one might anticipate a degree of romanticism. But his reading and interpretation of the documents seem unencumbered by sentiment. Consider his discussion of the motivation for migration to the Natchez District:

> Although people migrated to other regions of North America seeking economic opportunity, religious liberty, or the right to govern themselves after their own fashion, the early settlers of Mississippi moved there for no other reason than to make their fortunes. They were jealous of their freedoms and determined to conduct their affairs as they chose, but were not primarily concerned with protecting any particular culture. They valued their "peculiar institution" and their staple crop, not on philosophical grounds, but because slavery and cotton were profitable under Mississippi's combination of geography, soil and climate. As they unblushingly admitted, their foremost motive was the desire to acquire wealth. Thus, throughout the ante-bellum period, the economic factor was mainly responsible for shaping the social and political life—and to a lesser extent the intellectual life—of the state. (9)

While Moore himself did not employ Marxist interpretations, one can certainly see how he provided grist for their mill.

Reviews in history journals were favorable. Fellow historian of Old South agriculture Weymouth Tyree Jordan, writing in *Agricultural History*, classed the book with Cornelius O. Cathey's *Agricultural Developments in North Carolina, 1783–1860*, noting that as a work "based on fact rather than fancy, it is an admirable study." Jordan agreed with Moore's presentation that "cotton planters were engaged in a capitalistic undertaking" and that they "supported slavery because it helped them make money." Jordan contrasted Moore's view that slavery was often highly profitable with Sydnor's view that slavery in Mississippi was unprofitable.

Charles Sackett Sydnor had published *Slavery in Mississippi* in 1933. When Jordan wrote, "Reaching such conclusions, he [Moore] is more icon-oclastic than Sydnor," Jordan seemed to indicate that the twenty-five-year-old interpretation by Sydnor had attained the status of received wisdom in the history profession.[2] Chester McArthur Destler was also quite laudatory in his review in the *Journal of Southern History*, and he especially endorsed Moore's portrayal of the profitable agriculture of the 1850s over Lewis C. Gray's depiction of "inefficient, profitless southern farming."[3]

Agriculture in Ante-Bellum Mississippi appeared as historians were producing major studies of slavery. Cliometricians were interested in questions about the self-sufficiency of the plantation. Most economic his-torians accepted that slavery was profitable as a bottom line proposition, but they had many questions about the efficiency of resource allocation. Professor Moore had much to say, but decidedly not in a direct manner, about the controversies surrounding the nature of slavery. There are no historiographical essays or didactic digressions on the subject to interrupt the chronological and topical narrative. This is partly a stylistic preference. His interpretations were of a qualitative nature. An examination of his-tories written in the 1960s and 1970s leaves the impression that while appreciated by some historians, Moore's contributions were nonetheless underutilized in the writing of history. Eugene D. Genovese in *The Polit-ical Economy of Slavery* complimented the state studies of antebellum agriculture by Moore, Bonner, Jordan, and Cathey and lamented the lack of studies for other states. He also produced an article on the agricultural histories of the antebellum South.[4] Gavin Wright's *The Political Economy of the Cotton South* gave a more central role to the traditional historical literature than previous economic analyses. He sought to combine the two methodologies in his investigation of the cotton South.[5] It must be noted, however, that Robert William Fogel and Stanley L. Engerman did not find the Moore volume valuable for the writing of *Time on the Cross*, or per-haps they did not find it at all.

Moore's *The Emergence of the Cotton Kingdom in the Old Southwest*, published by Louisiana State University Press in 1988, covered the same temporal and spatial territory as the earlier volume. It differed on two counts—in what it added and in what it omitted. In the later volume, Moore devoted chapters to subjects such as transportation, towns and vil-lages, and manufacturing that had been examined only in the context of agricultural history in the earlier volume. And in his first book, Moore had stated that Charles Sackett Sydnor's *Slavery in Mississippi*, "made further research on this topic unnecessary" (10); while the later volume contained chapters on agricultural slavery and urban slavery, as well as chapters on rural whites and urban whites. *Emergence* also summarizes details found in the agricultural history volume. For instance the fascinating five-page story in his first book of the indigenous invention of the saw-toothed circular gin blade and the building of early saw gins was compressed into a few sentences in *The Emergence of the Cotton Kingdom in the Old Southwest.*

In 1970 Moore moved to the history department of Florida State University at Tallahassee to replace Weymouth Tyree Jordan, who had died at the all too young age of fifty-six in November 1968. Jordan had written *Hugh Davis and His Alabama Plantation* (University of Alabama Press, 1948, 1974), *Ante-Bellum Alabama: Town and Country* (Florida State University Press, 1957; University of Alabama Press, 1987), and *George Washington Campbell of Tennessee, Western Statesman* (Florida State University Press, 1955). The department wanted an established scholar to take the Old South teaching position while William W. Rogers continued teaching the New South course. It was probably a happenstance that both Jordan and Moore had specialized in agricultural history, but there are other similarities between the two. While specializing in a subject, agricultural history, that has never been wildly popular with presses, both produced solid works of history that merited being reprinted. Both also served as president of the Agricultural History Society. Moore retired from the history department at Florida State University in 1993. He and his historian wife, Margaret DesChamps Moore, edited the journal of one of her ancestors that was published in 1997 as *Cotton Culture on the South Carolina Frontier: Journal of John Baxter Fraser, 1804–1807.* Margaret DesChamps Moore died on December 7, 2003.

As memories of the initial publication of *Agriculture in Ante-Bellum Mississippi* recede into the past, it seems that more recent scholarship could use an infusion of its old lessons. Malcolm J. Rohrbough's *Trans-Appalachian Frontier,* for instance, repeats the fiction that "the new Cotton Kingdom spread across the South from western Georgia, across the landscape of Indian nations to Natchez."[6] A positive recent development was the use of Moore's work on cotton breeding in an article by Alan L. Olmstead and Paul W. Rhode, "Biological Innovation and Productivity Growth in the Antebellum Cotton Economy," in the *Journal of Economic History.*[7] This article confirms the criticality of understanding technology when examining productivity questions in economic analysis.

The purpose of this reprint edition and of the others in the University of South Carolina Press's Southern Classics series is not solely to honor scholars but, more important, to make monographs available to this generation. Regardless of the use of information from this book in more recent studies, today's students should read *Agriculture in Ante-Bellum Mississippi* unfiltered. More manuscript collections are available than in the 1950s when the volume was researched. Newspapers are more readily available, and a generation of economic historians has brought its skills to the task of understanding the past. Nonetheless much of Moore's narrative and his interpretations stand up well. And as a lagniappe he presents the story in a clear, crisp style.

Notes

1. John Hebron Moore, *Agriculture in Ante-Bellum Mississippi* (New York: Bookman Associates, 1958; New York: Octagon Books, 1971; Columbia: University of South Carolina Press, 2010). References hereafter will be made parenthetically in text and are to the 2010 edition.

2. Weymouth T. Jordan, review of *Agriculture in Ante-Bellum Mississippi*, by John Hebron Moore, *Agricultural History* 33 (April 1959): 90.

3. Chester McArthur Destler, review of *Agriculture in Ante-Bellum Mississippi*, by John Hebron Moore, *Journal of Southern History* 25 (February 1959): 121.

4. Eugene D. Genovese, "Recent Contributions to the Economic Historiography of the Slave South," *Science & Society* 24 (Winter 1960): 53–66.

5. Gavin Wright, *The Political Economy of the Cotton South: Households, Markets, and Wealth in the Nineteenth Century* (New York: W. W. Norton & Company, 1978), 1.

6. Malcolm J. Rohrbough, *Trans-Appalachian Frontier: People, Societies, and Institutions, 1775–1850*, 3rd ed. (Bloomington: Indiana University Press, 2008), 171.

7. Alan L. Olmstead and Paul W. Rhode, "Biological Innovation and Productivity Growth in the Antebellum Cotton Economy," *Journal of Economic History* 68 (December 2008): 1123–71.

To
CORINNE HEBRON MORRIS
and
DEAN HEBRON

Preface

Although people migrated to other regions of North America seeking economic opportunity, religious liberty, or the right to govern themselves after their own fashion, the early settlers of Mississippi moved there for no other reason than to make their fortunes. They were jealous of their freedoms and determined to conduct their affairs as they chose, but were not primarily concerned with protecting any particular culture. They valued their "peculiar institution" and their staple crop, not on philosophical grounds, but because slavery and cotton were profitable under Mississippi's combination of geography, soil and climate. As they unblushingly admitted, their foremost motive was the desire to acquire wealth. Thus, throughout the ante-bellum period, the economic factor was mainly responsible for shaping the social and political life—and to a lesser extent the intellectual life—of the state.

Despite the paramount influence of economic forces, historians who have written about the Southwest in the twentieth century have, with a few notable exceptions, neglected them— a mistake avoided, incidentally, by such contemporary Mississippi writers as Benjamin L. C. Wailes, John F. H. Claiborne and Horace S. Fulkerson. For this common oversight, the Civil War is principally responsible. The stark drama of that conflict so captured the imagination of succeeding generations of scholars that they tended to confine their study of the Old South to the causes of the war. The institution of slavery and the political struggle culminating in secession have long held the center of the historical stage, and only recently have economic and cultural aspects of Southern society come under serious investigation. Consequently, our knowledge of pre-war Mississippi is fragmentary and distorted.

9

This is especially true of the state's economic history. Adequate studies of transportation, finance, manufacturing, and even agriculture are lacking, although farming provided a vast majority of the population with their livelihood. Two comparatively recent monographs, however, have thrown much light on the state's economic organization. Charles Sackett Sydnor's definitive *Slavery in Mississippi* (New York, 1933) clarified the place of the Negro in ante-bellum agricultural history and made further research on this topic unnecessary. Herbert Weaver's *Mississippi Farmers, 1850–1860* (Nashville, 1945), a brief sketch of agriculture on the eve of the Civil War drawn from manuscript census schedules, is also of great value. Still lacking, however, is information on the transition from frontier farming to the complex agricultural system described by Professor Weaver. The purpose of this study is to bridge that historical gap by tracing the technological development of farming in Mississippi and analyzing the forces that produced the change.

The economic history of Mississippi, more than that of any other state, is the history of cotton. In the Cotton Kingdom, Mississippi was the *Ilê de France*—the heartland and the citadel—and its citizens were the most fiercely loyal subjects of the Crown. In Louisiana, planters grew sugar as well as cotton; in Tennessee, tobacco as well as cotton. In South Carolina and Georgia, as the ante-bellum era drew to a close, the people were turning in increasing numbers to animal husbandry, horticulture, and diversification of crops. Some of the citizens of South Carolina, Georgia and Alabama were even deserting their fields to take jobs in mills and factories. Not so the people of Mississippi! They chose instead to devote their energies to perfecting their type of commercial agriculture. Mississippians expanded their output of cotton and reduced their operating costs by breeding improved strains of it, by developing and adopting more effective farm implements and machinery, by improving their techniques of soil conservation, and by feeding and clothing their workers with the products of their own farms and plantations.

As a consequence of these improvements in agricultural methods, Mississippi became the leading producer of cotton during the decade of the 'fifties. With cotton prices unusually high during this period, the state's prosperity became a by-word throughout the South. This prosperity, however, was illusory; never again after 1860 was Mississippi to enjoy a position of such eminence in the economic life of the nation. Events soon demonstrated that the cotton bale alone, however efficiently produced, was not a sound foundation for the economy of a state.

This study owes much to the work of Professors James C. Bonner, of Georgia State College for Women, and James Z. Rabun, of Emory University. The impulse to investigate agriculture in ante-bellum Mississippi originated in an Old South seminar conducted by Professor Bonner, and execution of the project was made easier by his advice and encouragement and by reliance upon his path-breaking work in ante-bellum agricultural history. Any merits in style and organization this study may possess can be attributed to the pencil of Professor Rabun. Any errors of fact or interpretation, however, I claim as my own.

I owe special appreciation to Dr. William McCain and Miss Charlotte Capers for the generous manner in which they placed the facilities of the Mississippi State Department of Archives and History at my disposal, and also to the editors of *Agricultural History* for permission to republish portions of two chapters which appeared earlier in their journal.

University of Mississippi   J. H. M.
June 1, 1957

CHAPTER I

Mississippi's Search for a Staple Crop

Introduction of Eli Whitney's newly invented cotton gin into the Natchez region and the signing of the Treaty of San Lorenzo made the year 1795 a major turning point in the political and economic development of early Mississippi. Both events were far-reaching in their consequences. The Treaty of San Lorenzo transferred title of the Natchez District of the old province of West Florida from Spain to the United States by locating the boundary line separating the territories of these two countries at the thirty-first parallel. What is even more important, Whitney's gin placed the inhabited area of Mississippi upon a much sounder economic footing than it had ever before enjoyed, for it reduced costs of raising cotton to the point where the white staple soon became the standard crop of that section of the Old Southwest.

The Natchez District of West Florida experienced little of either peace or prosperity before 1795. Within the span of a single generation, flags of three major European powers floated over its seat of government. France, the original owner, after an occupancy of more than half a century, was forced to cede that region to Great Britain in 1763 as a result of defeat in the Seven Year's War, and Britain in turn lost it to Spain during the War for American Independence. Under political conditions as disturbed as these, there was small opportunity for a young frontier colony to grow and prosper.

During the long period of French domination, almost nothing was done toward occupying the rich lands along the lower east bank of the Mississippi River. True, between 1716 and 1729 a few hundred farmers with their families were settled in the vi-

13

cinity of Fort Rosalie, the present site of the city of Natchez.
Yet these immigrants were either killed or driven away during
an Indian uprising in 1729. The French quelled that insurrection
but made no further efforts to bring in settlers interested in farming. With their main endeavor focused upon the Indian fur trade,
the French devoted so little attention to agriculture during the
remainder of their stay in the Lower Mississippi Valley that the
British found few signs of their earlier presence upon their arrival at Fort Rosalie. Except for the ruins of the fort itself and
a few clearings hard by, no man could have told that white
settlers had ever dwelt there.[1]

The first permanent agricultural settlements in the territory
included within the present state of Mississippi were made by
Englishmen. After securing West Florida from the French, the
ministry in London adopted a policy of encouraging a flow of
immigration into this border region. It made generous land grants
to war veterans and to political friends of the regime. When the
area previously cleared of Indians by the French proved to be
insufficient to satisfy the demand, the Superintendent of Indian
Affairs in 1777 bought from the Choctaws a strip of territory
running north along the Mississippi River from Loftus Heights
to the mouth of the Yazoo, a distance slightly in excess of one
hundred miles.[2]

In spite of the British government's friendly attitude toward
prospective settlers in West Florida and notwithstanding the
availability of fabulously rich alluvial soils along the lower Mississippi River and its tributaries, comparatively few persons
were attracted to the Natchez District during the period of British domination. Like Virginia nearly two centuries earlier, West
Florida did not thrive until its settlers began to shift their labors
away from subsistence farming and to cultivate, instead, crops
that could be exported to European markets. This step was not
taken under the British flag. Nevertheless, the population of the
Natchez District did grow steadily, albeit slowly, after 1770.[3]
During the American Revolution an influx of American loyalists
temporarily speeded the slow expansion of the colony. As many
of these newcomers were persons of wealth and energy, they
added materially to the prosperity of the Natchez District.[4]

Although the residents of West Florida attempted to remain neutral during the American Revolution, they were not able to avoid being drawn into the conflict. The Natchez District first was raided by American irregulars in 1778 and then occupied permanently by troops from Spanish Louisiana in the following year. The British garrisons stationed in widely scattered posts were so small and weak that they stood almost as an invitation to a resolute enemy to invade the colony. Don Bernardo de Galvez, the Governor-General of Louisiana, was not a man to allow such a prize to slip between his fingers through lack of initiative. Gathering together a formidable body of troops, Galvez opened hostilities against British Florida in 1779. By seizing the forts at Natchez, Manchac and Baton Rouge he obtained possession of all English territory along the Mississippi River between the mouth of the Yazoo and the city of New Orleans. The Gulf Coast forts fell into Spanish hands almost as easily as had those along the river. Mobile capitulated in 1780, followed in 1781 by Pensacola, the principal British stronghold in Florida. By the end of the War of American Independence, all of Florida was under the flag of Spain, and Spanish troops were distributed in fortified positions throughout the colony from Walnut Hills, just south of the outlet of the Yazoo, to the Atlantic seaboard.[5]

Notwithstanding Spain's wartime conquest and occupation of West Florida, Great Britain managed to make use of that province during the diplomatic maneuverings that ended the conflict. In order to cause future difficulties between the United States and Spain, the British negotiators in 1783 led each of these two powers to believe that it was acquiring the territory situated between the thirty-first parallel and a line running through the mouth of the Yazoo River. In the case of Spain, her claim to these lands arising from her interpretation of the Treaty of Paris was re-enforced both by the right of military conquest and by actual occupation of the area in dispute. The United States, in contrast, had no more than a shadowy claim based solely upon a supposed cession by Great Britain of territory which that country did not occupy at the time of the peace negotiations. This claim derived some legal substance from the fact that the British Crown, in a commission to the governor of Georgia in 1764, defined Georgia's southern boundary as the thirty-first parallel from the Mississippi to the Chattahoochee (and thence down the Chattahoochee to a line drawn from the St. Mary's). The United States over the next twelve years continued to insist that Spain surrender her holdings north of the thirty-first parallel and east of the Mississippi River. After doggedly resisting American diplomatic pressure, Spain finally acquiesced to the repeated American demands in 1795, agreeing in the Treaty of San Lorenzo to accept the thirty-first parallel as the northernmost boundary of her province of West Florida. As the District of Natchez was within the region affected by the treaty, that part of Mississippi thus ceased to be a pawn in the game of European power politics, although final transfer of authority from Spain to the United States was not accomplished until 1798, three years after approval of the treaty.[6]

Undeniably, the uneasy political situation prevailing in the Lower Mississippi Valley during the eighteenth century had made a climate unfavorable for development of the Natchez District. Yet that region had received some tangible benefits from each successive change of ownership. The French in their war with the Natchez tribe in 1729 took the first step toward establishing a colony by driving the original Indian inhabitants out of a considerable expanse of territory; but the France of that era had neither the interest in agriculture nor the surplus population required to set up a permanent self-supporting colony upon the soils thus summarily opened up to white settlement. England, on the other hand, was able to furnish the people needed for colonization when she acquired possession of the region around Fort Rosalie. She, however, could not provide her colonists with a nearby market for their produce. New Orleans by then was in the hands of Spain and closed to trade with British subjects. Consequently, immigrants coming into West Florida from the British Isles and the eastern American colonies were forced to depend mainly upon subsistence farming for their livelihood, exporting to Great Britain and the islands of the West Indies only very limited quantities of such products as cotton, corn, indigo, tobacco and barrel staves.[7]

When the District of Natchez passed into the possession of

Spain in 1779, its economic situation changed temporarily for the better. Natchez produce was admitted freely into the New Orleans market, and the Madrid government took pains to foster the economic well-being—and therefore the loyalty—of His Catholic Majesty's newly acquired English-speaking subjects.

Considerations of grand strategy were responsible for the Spanish government's unusually conciliatory attitude toward the non-Catholic population of the Natchez District. While the province of West Florida then possessed little economic value, its geographic position made it an important link in a chain of defenses guarding New Orleans, Louisiana and Mexico. The authorities in Madrid wished to establish a buffer zone between Louisiana and the Americans in Georgia and Tennessee. They believed that the gravest peril threatening the security of New Spain was the westward expansion of the United States. The Spaniards planned to safeguard the overland route into the Lower Mississippi Valley by welding the Choctaw, Creek and Chickasaw tribes into fighting forces sufficiently strong to resist American penetration into their respective territories. The northern river approach to the city of New Orleans they would defend by a prosperous, populous and loyal colony to be constructed out of the old Natchez District of West Florida.[8]

In carrying out the latter project, the Spaniards in Louisiana were hampered seriously by the fact that neither Old nor New Spain could provide settlers for Natchez in sizable numbers. If the population of that colony were to be enlarged as planned, it would be necessary to admit persons of nationalities other than Spanish and of religions other than Catholic, a step that was inconsistent with Spain's traditional colonial policy. Nevertheless, a high level decision was made in 1788 to welcome immigration into Louisiana and West Florida from the protestant republican United States.[9] In accordance with this revolutionary approach to the colonial problem, newcomers to the Natchez District were granted lands on more liberal terms than the British had given. They were exempted both from imperial taxation and from all obligations involving military service, and they were allowed to preserve their own religion provided that non-Catholic services should be held only in private homes.[10]

In the Natchez District this novel policy of conciliation was carried to even greater extremes than in the remainder of Spanish Louisiana. In that district most governmental offices were filled by local citizens who were usually of British descent. No attempt was made by the Crown to interfere with the property, language or customs of the inhabitants. Moreover, the Governor-General took upon himself the task of strengthening the economic position of the Natchez farmers by purchasing their tobacco at prices higher than those offered in the open market. This program of government aid tended to destroy the region's self-sufficiency by encouraging dependence upon a single staple crop.[11]

For a time the new colonial policy produced the very results in Natchez that the Spanish government hoped for. Numerous settlers, attracted by high tobacco prices, moved into the District, and many of the older inhabitants expanded the amount of land they cultivated. Between 1787 and 1792, the white population of the District increased from 1,926 persons to 4,300, the Negro population trebled, and the annual production of tobacco climbed to an estimated half a million pounds.[12]

In the Natchez region this period of prosperity and expansion was of brief duration. The Spanish government in 1790 was compelled to abandon its expensive price support program by new strategic considerations. Planning to split the American West off from the remainder of the United States by intrigue, and hoping to bind the farmers of that vast frontier section to New Spain with economic ties, the Spaniards in that year threw open the market at New Orleans to western produce—including tobacco— allowing it to enter Spanish territory subject to only a nominal duty.[13] Simultaneously, the government warehouses ceased to purchase Natchez tobacco at the old premium price. As the government thus was no longer buying largely on the open market, prices of all commodities except cotton dropped in 1792 to levels less than half those of 1788.[14]

This collapse of the New Orleans tobacco market was a major disaster to Natchez farmers. The government's price-fixing policy had encouraged heavy borrowing by the planters for expansion of tobacco growing, and as a class they were not prepared to face a sudden loss of their crop subsidy.[15] The plight of even the largest planters soon became so desperate that Estevan Miro, Governor-General of Louisiana, was compelled to intervene for them in order to prevent seizure of their property by New Orleans merchants. Nevertheless, the stay law issued by Miro in Louisiana and approved by the authorities in Madrid served only to postpone the day of reckoning; for it gave the ruined farmers an eight-year period of grace in which to pay their creditors.[16] It did nothing to resolve their fundamental difficulty—lack of a market for their only staple crop, tobacco. On this point the Spanish government was adamant: it would no longer undertake to sustain the unprofitable Natchez colony by purchasing its tobacco at artificially inflated prices.

Discontinuing the raising of tobacco for other than domestic use, farmers of the Natchez region in 1793 and 1794 concentrated their attention upon developing indigo into a crop that could take the place of tobacco as the mainstay of the region's economy.[17] After a promising beginning, this second attempt at finding a staple also ended in failure. In 1794 insects damaged the Mississippi indigo crop so severely that it was not planted again.[18] With this last fiasco the economic future of the debt-ridden Natchez region seemed dark indeed.

The people of early Mississippi finally discovered a solution for their persistent economic problems through chance rather than design. Recent experiments with tobacco and indigo had brought the entire District close to the brink of financial disaster. The only cash crop available to them was cotton, inasmuch as Mississippi pork, beef and corn had been virtually driven out of the New Orleans market by produce from Ohio, Kentucky and Tennessee. Thus, because they saw no other course open to them, farmers of Mississippi turned in 1795 to the cultivation of an upland variety of long staple black seed cotton as their principal money crop.[19]

If extensive cultivation of cotton had been attempted even as late as 1794, the experiment doubtless would have ended in failure. Their experience with raising cotton in limited quantities over a period of many years had demonstrated conclusively to farmers that the staple was not suited to conditions then existing in the Lower Mississippi Valley. It was too expensive to cultivate, to gather and to prepare for market, even at the price level prevailing during the 1790s, high though that price might have seemed to later growers in the 1800s. In 1795, however, the situation changed. The cost of preparing the cotton for market was sharply reduced by introduction of a new technological discovery into Southern agriculture, Eli Whitney's famous cotton engine, or gin. The rotating saws of the machine Whitney patented in 1794 could separate cotton fiber from the seeds cheaply and easily. With it, citizens of the Natchez District found they could grow cotton with assured profits.

Cleaning cotton for market had been a slow, tedious and costly process before the invention of Whitney's gin. In the Lower Mississippi Valley, seed had been removed from the lint either by hand or by means of a primitive roller gin modeled after a device employed in India for many centuries.[20] Machines of this type used in eighteenth century Mississippi had been quite simple in principle and construction, but very inefficient in operation. Two small wooden cylinders set closely together within a framework mounted upon a bench were revolved in opposite directions by means of a hand crank or foot treadle. When seed cotton was brought into contact with these rotating cylinders, lint was wrenched from the seeds, which were too large to pass through the narrow aperture, and drawn by friction between the rollers, emerging free of seed on the opposite side of the machine. Two persons were required to keep a friction or roller gin in continuous motion. One fed raw cotton into the rollers, the other turned the crank and received the ejected lint. The daily output of hand operated roller gins was discouragingly small. Under ideal working conditions one machine could clean at most seventy-five pounds of fiber a day, a rate that made the cost of preparing cotton for market almost prohibitive.[21]

When the first machine built upon Whitney's principle was given a trial in Mississippi in the fall of 1795, it demonstrated that it could clean the long staple black seed cotton of the Lower Mississippi Valley fully as well as the short staple green seed variety of Georgia for which it had been designed. On the black seed cotton the saw gin was vastly more efficient than the roller gins then in use. It could clean five hundred pounds of lint daily with no more labor than the roller gin required to produce a tenth as much.[22]

The importance of a gin of this remarkable degree of efficiency to Southwestern farmers could hardly be over-estimated. Cleaned cotton at the time was selling for twenty-five dollars a hundred pounds, while seed cotton brought no more than four dollars a hundred. Thus ginning at least doubled the value of raw cotton.[23] The Whitney saw gin, in fact, was to farmers of Mississippi what the fabled philosopher's stone had been to medieval alchemists—a catalyst transforming base metal into gold.

Whitney's gin was a radical departure in principle from all of its predecessors. Where the latter had employed friction rollers to separate lint from seed, the former made use of rotating wire fingers to draw fiber through slits too narrow to permit the passage of even the very small seeds of Georgia upland cotton. The lint, after passing through the grate, as the removable slitted side of the hopper was called, was swept from the wires by brushes mounted upon a rapidly revolving flywheel. The wind created by this fan-like device blew the fiber into an almost airtight compartment provided for its reception. The seeds which had been trapped on the other side of the grate dropped to the bottom of the hopper containing the unginned cotton and from there fell to the ground through a spout.[24]

The saw gin which Whitney demonstrated in Georgia as a working model in 1793 and as a full-sized machine in 1795 was so simple in construction that it could be duplicated by a competent workman using only hand tools to be found on all plantations in the Old Southwest. Power for operating Whitney saw gins could be obtained from any source supplying rotary motion. At different times hand cranks, "horsepowers" of various kinds, water wheels and steam engines were used with satisfactory results.[25]

Mississippi's first saw gin was constructed in violation of Whitney's patent rights during the summer of 1795 and put into operation in September.[26] Daniel Clark, Sr., a wealthy planter of Wilkinson County, designed this famous machine after examining drawings made by a traveler who had seen one of Whitney's gins while on a trip to Georgia.[27] Clark's gin was constructed by local mechanics working under his supervision on his Sligo plantation situated near Fort Adams. Although witnesses years after the event differed as to the identities of the workmen Clark employed, it appears likely that much of the work was done by James Bolls, Jr., son of a Scottish plantation owner of the same name, with the aid of Barclay, a skilled Negro slave belonging to Clark, and a local blacksmith named Hughes.[28] Clark, being unable to obtain satisfactory iron wire of the type used by Whitney, was forced to substitute circular iron saws cut by hand from hoe blades and filed into shape.[29] After tests, this innovation proved to be so successful that Clark's saws rather than Whitney's wires were adopted as standard for all gins constructed subsequently in Mississippi, though not in other states.

Clark's gin at Sligo attracted a great deal of attention from planters of the Natchez area even while it was in the process of construction, and many of them travelled to Fort Adams from considerable distances in order to be present at some of the early tests. When these trials clearly demonstrated that saw gins were vastly superior to the best of roller gins, several other planters of the wealthier class had saw gins of the Clark variety built upon their plantations. Among the earliest to take this step were William Dunbar, a Scottish scientist and correspondent of Thomas Jefferson; Colonel Anthony Hutchins, a retired officer of the British army; and William Voursdan, son-in-law to Hutchins and the first planter to ship cotton directly to England.[30]

Because the demand for saw gins became very great during the next few years a number of local mechanics embarked upon careers as professional ginwrights. The first of these to become well-known in Mississippi was James Bolls, Jr., who had gained experience at this new trade by working on the Sligo gin. After completing his work for Clark, Bolls next was employed by William Voursdan to construct a gin for his plantation, "Cotton Fields," situated between Natchez and the town of Washington, and after that by the merchant Robert Mason to erect a public toll gin in the city of Natchez.[31] Among several other gins built before 1800 was one set up by Bolls upon the plantation of his father, James Bolls, Sr.[22] David Greenleaf, who settled in Natchez in 1795, was another mechanic who gained a reputation in Mississippi as a pioneer manufacturer of saw gins and other machines used upon Southwestern cotton plantations. He built the state's first public gin in 1796 upon land belonging to Richard Curtiss, near Selsertown. After making a number of other gins for private use, he built one in 1798 to be operated as a commercial enterprise by himself in partnership with two Scottish merchants, David Ferguson and Melling Worthy.[33] In addition to his work with early gins, Greenleaf also appears to have originated the practice of packing cotton in a press equipped with screws turned by hand. It is certain that he constructed the first model of this type of machine to be used in Mississippi before 1799.[34]

Eleazer Carver was the most successful of the early ginwrights. After entering the business at Washington in 1807, he made the first important improvement in the design of Southwestern gins since Clark replaced Whitney's wires with circular cotton saws. Carver improved the grates by changing their shape and thus reduced their tendency to become clogged with lint while the machine was in operation. Carver's improved gins became very popular with planters of the Old Southwest, and his business grew to such dimensions that he was compelled to erect Mississippi's first saw mill in order to assure himself of an adequate supply of lumber. In later years Carver transferred his operations from Mississippi to Bridgewater, Massachusetts, to take advantage of New England's superior supply of skilled workmen and its recent improvements in the design and manufacture of machine tools. The Carver company prospered in its new location and by the 1850s had become one of the largest producers of cotton gins in the United States.[35]

Many of the new type of cotton gins were manufactured and put into use in the Natchez District in the period between 1795 and 1800. By the latter date one could be found upon almost every large plantation.[36] These machines were all constructed in the same manner and upon the same principle, but they differed from one another in their means of motive power and in their size and rate of production. The metal parts of all of them were fabricated in local blacksmith shops. As there were neither forges nor foundries in that region as yet, the saws and other metal parts were cut from sheet iron or bar stock with hand tools and then hammered and filed into shape.[37] Cotton saws made in this fashion were particularly expensive, ordinarily costing as much as five dollars each.[38] The woodwork of the ginstands was erected on the spot by carpenters who used imported lumber or hand-sawed boards because there were no saw mills in the District before 1810. Some few of these early gins were driven by water power, but the high cost of building dams and the difficulty in keeping them in good condition in this region of rock-free soils caused a vast majority of gin owners to resort to the use of horses and mules for driving their machinery.[39] Daily output of these Mississippi gins at the turn of the century averaged between five hundred and fifteen hundred pounds of cotton lint.[40]

Small farmers who were unable to afford gins of their own transported their seed cotton to commercial establishments or to plantations of wealthier neighbors. There it was cleaned for them at a toll, usually amounting to ten percent of the cotton deposited at the gin.[41] As roads were almost non-existent and carts and wagons were rare, seed cotton from the interior was often packed in bags and carried on horseback to the gin over distances sometimes as great as twenty miles.[42]

Cotton production in Mississippi increased very rapidly during the last five years of the eighteenth century, and gins in the Natchez region were unable to cope with the demands made upon their services. Many privately owned machines on plantations were pressed into public service along with the ones belonging to commercial enterprises. Even then the facilities could not process all the crop in a year's time. Consequently, farmers often had to wait many months after leaving their cotton at a gin before obtaining delivery of their lint. As this delay became more pronounced, the practice of selling cotton while it was still in the possession of the ginner became the general rule in the District. Ginners' cotton receipts were accepted freely by merchants in payment for debts at a standard rate of five cents a pound for unginned cotton, and these negotiable certificates for a time assumed the aspect of a circulating medium of exchange.[44] Their use as local currency became so common during the early 1800s that they were recognized as legal tender by the territorial government of Mississippi and regulated as such by law.[45]

Large scale cultivation of cotton for export to foreign markets by 1799 had passed beyond the experimental stage and was generally recognized as the principal basis of agriculture by farmers cultivating the rich lands along the lower Mississippi River.[46] Experience had demonstrated that it could be cultivated with slave labor as easily as tobacco and much more easily than indigo. In all ways, according to William Dunbar, a planter who shipped more than one hundred bales (averaging three hundred and thirty pounds each) to Liverpool in 1800, cotton was an unqualified boon to the region. This view he expressed to his partner, John Ross of Philadelphia, in a letter dated May 23, 1799, in the following words:

> It is by far the most profitable crop we have ever undertaken in this country. The climate and soil suit it exactly, and I am of opinion that the fibre, already of so fine in [sic] quality, will be still better when our lands are well cleared and the soil properly triturated. The introduction of the rag-wheel gin was fortunate indeed for this district. I have reason to think that the new gin has been greatly improved here. Our latest and best make, injure the staple little more than cards.[47]

In Dunbar's opinion cotton had proved itself to be the ideal staple crop for the Mississippi Territory, and his belief was shared in all respects by a multitude of planters and farmers who had taken up the cultivation of cotton as their major crop.

With the fertile lands of the old Natchez District producing average yields of fifteen hundred to two thousand pounds of seed cotton to the acre and approximately eight hundred pounds to the farm worker at prices ranging between twenty and twenty-five cents a pound for the fiber, this new staple crop was unquestionably returning handsome profits to those engaged in its cultivation. In 1801 the cotton crop alone earned some seven hundred thousand dollars for farmers of the District, a sum that represented an average income in excess of seven hundred dollars for each of the region's nine thousand inhabitants.[48] Thus cotton finally brought a most unaccustomed prosperity to the people of the new Territory of Mississippi, and put an end to their long search for a farm product profitable enough to render their part of the country economically self-supporting.

CHAPTER II

Evolution of Petit Gulf Cotton

While in its initial phase of development, Mississippi's agriculture underwent a radical transformation. Farmers of this region abandoned all pretense of economic self-sufficiency and replaced their traditional subsistence farming with the cultivation of cotton for export. By the close of the eighteenth century the economic transition had been completed, and cotton had gained universal acceptance as the staple crop.

In the second stage of agricultural development lasting from 1800 to 1837, Mississippi farmers devoted their attention to the multiple task of improving methods of cotton cultivation, perfecting the tools, implements and machines needed to grow and prepare cotton for market, and developing varieties suited to local farming conditions. By attaining at least limited success in each of these different fields of endeavor, the pioneer cotton growers of the Old Southwest made their occupation profitable enough to attract ambitious immigrants to Mississippi from all parts of the Union. Upon arrival the newcomers took up the culture of cotton, acquired the techniques of cultivation developed in the old Natchez District, and spread them throughout the state.

By 1830 the plantation system had taken root in the southern half of Mississippi wherever there was adequate water transportation and soil fertile enough for cotton. By far the greater number of cotton plantations were located in the "river counties" of the old Natchez District, and in the region surrounding the present town of Columbus, which the Chickasaws surrendered in

27

1816. Settlement of North Mississippi began in earnest when the lands ceded by the Choctaws and Chickasaws in the Treaties of Pontotoc and Dancing Rabbit Creek were placed on the market in 1833.

During the period of change and development ending with the depression of 1837–49. Mississippians made a contribution to cotton growing which was second in importance only to the invention of Eli Whitney's cotton gin. This innovation was a newly developed strain of cotton, created by crossing several varieties of the plant grown in the New World by the French, British and Spanish respectively. Subsequent long cultivation and selective breeding adapted this new cotton to a wide variety of soils and brought it to a near perfection in quality. Because of its numerous valuable properties, the so-called Mexican cotton of Mississippi was carried from its original home in the 1830s and introduced into all of the cotton-growing states. Being the earliest Southern standard variety, Mexican became the ancestor of all modern American breeds.

The cotton which was raised in the Lower Mississippi Valley at the beginning of the nineteenth century was quite different in origin and general characteristics from that common to the states of the Atlantic seaboard. Georgia Upland or Green Seed cotton (*Gossypium Hirsutum, Linn.*) as the eastern strain was called, was first cultivated in North America in 1734. In that year experiments were carried out in Georgia with seeds of this distinctive variety obtained from the horticultural garden of a famous English botanist, Philip Miller of Chelsea. Although the original home of the Green Seed has never been ascertained, Miller probably obtained his original specimens from the island of Guadeloupe in the West Indies. The hardy Green Seed cotton was found to be suited to the climate and soils of the Georgia uplands, and from there its cultivation spread rapidly into similar topographical areas in the Carolinas, Virginia and Tennessee. Hence by 1786 Georgia Upland had almost wholly replaced other types in the interior regions of the Eastern United States. It was not immediately introduced into the Lower Mississippi Valley, however; for another variety of cotton with a greater commercial value was already being raised there.[1]

Georgia Upland of the early 1800s was a short staple cotton most easily distinguished from other varieties by its seeds, which were small, fuzzy and green in color. The low bush-like stalk had many branches, and its leaves, leaf stems and veins were coated with a profusion of hairs that gave to the Green Seed cotton the botanical name of *hirsutum*. Its lint was short and coarse and very hard to separate from the woolly green seeds.[2] Until the invention of Whitney's saw gin, Georgia Upland was prepared for market by wrenching the fiber from the seed by hand or by the taut string of a wooden bow, a primitive method of ginning that made this variety of cotton known to the European textile trade under the name of Georgia Bowed cotton.[3] Because of the poor quality of its staple the Green Seed brought a much lower price in European markets than did the long staple black seed types of the West Indies, the Lower Mississippi Valley, and the Georgia Sea Islands.[4]

Experiments intended to adapt the Sea Island cotton of the West Indies (*Gossypium Barbadense, Linn.*) to conditions existing along the lower reaches of the Mississippi River were conducted in Louisiana by the French during the early 1720s, but they were unsuccessful everywhere except near the Gulf of Mexico.[5] When the French colonists discovered that winters were often severe enough to kill the perennial West Indian cotton tree back to its roots before its lint matured, they discontinued cultivation of the Sea Island variety and began to plant another type that had been imported from Siam. On trial the new cotton proved to be successful. In 1733, it was reported that the Siamese Cotton (*Gossypium Nanking, Meyer.*) was the only variety then being grown in Louisiana. It was thriving wherever planted, and was producing abundantly with a minimum of cultivation.[6] Additional importations of the same strain were made during the 1750s, and it was planted thereafter with such success that it remained the standard variety of the Lower Mississippi Valley until about the year 1810.[7]

The Siamese breed, called Black Seed or Creole cotton by ante-bellum Mississippians, was so remarkably similar in many respects to Sea Island that the inhabitants of the Mississippi Valley soon forget its true place of origin and adopted the erroneous opinion that it had evolved from the arboreal West Indian cotton.[8] This was a natural mistake. A small quantity of the true Sea Island cotton was raised along the Gulf throughout the whole period between 1720 and 1860, and its outstanding characteristics were much like those of the Siamese cultivated further to the north. Both varieties had large smooth black seeds. Both produced fine pure-white lint of great length and high commercial value, and the seeds of both could be separated from the floss by means of the simple roller or friction gin. The Siamese cotton, in fact, differed from Sea Island in only a few important details. It was an annual plant instead of a perennial, its yield was greater than the Sea Island, and its seeds were more difficult to remove from the lint. Its staple also was somewhat shorter than that of the West Indian variety.[9]

Although the plants of the Siamese black seed cotton did not grow so large as the Sea Island cotton trees of the Gulf Coast, they nevertheless attained a size unusual among upland cottons when planted in the rich alluvial loams of the river districts of Mississippi. The branches of adjoining mature plants usually interlocked with one another, even though they were customarily planted four feet apart in drills spaced at intervals of four and one half feet. If the tops were not trimmed off at about the age of four months, the stalks would grow too tall to allow pickers to reach the uppermost bolls easily.[10]

Creole cotton was as productive as any variety cultivated in Mississippi before the Civil War. It yielded from fifteen hundred to two thousand pounds of seed cotton to the acre when planted in fresh soils of river bottoms, and about half as much when grown on fertile hill land. The lint was fine in quality. Its staple was about three quarters of an inch in length, and the fibers were very delicate and of unusually great tensile strength. The floss was ordinarily pure-white in color, but rich creams of the Nanking type were not unusual.[11] Because of these various factors, demand was good for the Creole cotton. During the years between 1795 and 1825, it commanded a price ranging between fifteen and thirty cents a pound, except for a short period during the War of 1812.[12]

To some extent the numerous good qualities of the Creole

strain were offset by a few less fortunate characteristics. The amount of salable lint derived from the seed cotton was disappointingly small because of the great size and weight of its distinctive black seed. In the process of ginning, the weight of this variety was reduced by seventy-five percent. When compared with the thirty-three percent of lint usual to Mississippi cottons of the 1850s, the Creole's proportion of lint to seed was not impressive. Moreover, black seed Creole cotton was not easy to pick. Its bolls, which were divided into three sections, were not much larger than pigeons' eggs, and their lint clung tenaciously to these small pods even after the cotton was fully ripe. Consequently, cotton pickers were able to gather less than one hundred pounds in a day's time. The difficulty in harvesting this cotton was so great, in fact, that it was not unusual for unpicked cotton still to be standing in the field in February at a time when preparations were going on for planting the next season's crop.[13]

Prior to 1811, the Black Seed upland cotton of the old Natchez District was remarkably free from the ravages of plant disease and insects.[14] True, the army worm made its first appearance in force in Mississippi in 1804 and, by stripping the leaves from immature plants, reduced the crop around Natchez by an estimated twenty-five percent.[15] After causing widespread alarm in the cotton growing community that year, the voracious worm then disappeared and was not heard of again until after the close of the War of 1812. In 1811, however, the Creole cotton of the Lower Mississippi Valley became infected generally by a plant disease known to farmers as the rot.[16] This bacterial disease, probably caused by the *Bacillus Gossypium Stedman,* attacked the cotton boll, often destroying its entire contents.[17] As the true cause of such diseases did not become known until studied under microscopes at a much later date, cotton growers of the early 1800s were unable to take any effective steps to combat the menace that was threatening their principal means of livelihood.[18]

The deadly rot continued to spread through Mississippi for fifteen years, and everywhere caused serious damage to the Creole cotton crop. In desperation many farmers began to plant the less susceptible though inferior Georgia Upland cotton in addition to their traditional long staple breed. Seeds of this Upland cotton were imported from the Cumberland region of Tennessee and from Central Georgia. The Green Seed cotton, however, never wholly displaced the more valuable Black Seed, and both varieties are known to have been cultivated on the same plantations as late as 1834. After its introduction into Mississippi in 1812, Georgia Upland was comparatively immune to the prevalent rot for a number of years before it too succumbed. In the mid-1820s, both the Creole and Georgia Upland varieties suffered extensive damage from an especially virulent outbreak of the dreaded plant disease. This disaster had the effect of causing Mississippi planters to recommence their search for strains of cotton possessing some degree of immunity to the rot.[19]

Having become accustomed to the superior qualities of the long staple Creole cotton, Mississippians would never have been satisfied with the short staple Green Seed even if it had remained free of rot. For when compared with Black Seed cotton, Georgia Upland was decidedly inferior in yield of seed cotton as well as in quality of lint. Because of its very small hard-shelled pods, Green Seed cotton was more difficult to gather than the Creole cotton, which was itself hard enough to pick. An extract from the cotton book of G. W. Lovelace, dated October, 1817, reveals that not one of Lovelace's slaves was able to pick as much as sixty-five pounds of the Green Seed cotton in a day, even at the height of the picking season in a crop grown upon fertile river lowlands.[20] The added fact that Georgia Upland was much more difficult to gin than the Black Seed variety makes it all the easier to understand why growers of Mississippi were eager to find a disease-resistant strain of cotton possessing some of the superior qualifications of the black seed Creole.

Fortunately, such a cotton was available close at hand. In the year 1806 Walter Burling of Natchez had obtained in Mexico some seeds of a cotton that had been cultivated for many centuries by the Indians of the central Mexican plateau and had smuggled them out of the country.[21] Burling, upon his return home, gave some of his Mexican seeds to William Dunbar, a close friend and neighbor. Dunbar cultivated them experimentally in 1807 along with some other seed of the tan-colored Nanking cotton, and at the end of the season sent specimens of their lint to Liverpool for examination by textile experts.[22] It is probable that the English experts sent Dunbar a favorable report on his Mexican cotton, for during the next three years he continued to expand his plantings of this variety. By the time of his death in 1810, the old scientist had succeeded in increasing his yearly output of the Mexican strain to more than three thousand pounds of ginned cotton.[23] In time, other planters of the neighborhood followed the example of Burling and Dunbar and took up the cultivation of the imported variety.

In 1820 a traveler from the Eastern states reported in an agricultural periodical that the Mexican cotton (*Gossypium Mexicanum, Tod.*) of the Lower Mississippi Valley had many characterestics valuable to cotton planters everywhere.[24] It ripened earlier in the fall than any other type then in cultivation in the United States, and it displayed a noticeable tendency to mature many of its bolls simultaneously. Even more important, it possessed exceptional picking properties. Its large four or five-sectioned bolls opened so widely upon ripening that their lint could be plucked from the pod more easily than could any other known variety of the staple. Because of this unusual quality, pickers could gather three to four times as much of the Mexican in a day as they could of the common Georgia Green Seed cotton and almost twice as much as Creole Black Seed cotton. Most important of all, the Mexican strain was totally immune to the rot.[25] In contrast to these valuable properties, the Mississippi Mexican cotton of the 1820s possessed only one serious fault. It could not be left unpicked in the fields after the bolls had begun to ripen. In this regard the very quality that made Mexican so valuable gave planters of that variety their greatest difficulty. The lint that was so easy to gather was likely to fall to the ground of its own accord shortly after the pods opened fully. Consequently, planters of the true Mexican variety of cotton were always in danger of suffering losses from early fall winds and rains.[26]

While increasing numbers of cotton growers in the Lower Mississippi Valley were taking up the cultivation of Burling's rot-resistant Mexican cotton in the 1820s, that strain was undergoing gradual changes. These modifications resulted in part from the adjustment of the plant to new conditions of soil and climate and partly from the accidental crossing of the imported Mexican strain with other varieties. This crossing of the Mexican with the Georgia Green Seed and possibly also with the Creole Black Seed ultimately produced a hybrid cotton possessing some of the more valuable properties of each of the parent stocks. The new breed retained the woolly white seeds, the large bolls, the good picking properties, and the immunity to the rot which were the distinctive features of Burling's Mexican cotton. From the Georgia Upland it inherited the ability to retain lint in the pod for a much longer time after ripening than the Mexican stock had possessed. In size the plants of the new breed were somewhere between the very large Mexican and the small bush-like Georgia Upland. Although the lint was somewhat shorter and coarser than the high grade long staple of Burling's cotton, it was nevertheless rated as an excellent specimen of short staple much superior in quality to the best of Georgia or Tennessee Green Seed.[27]

By 1830 the natural processes of crossbreeding and evolution had produced a new variety of cotton that came very close to meeting the minimum requirements of Southern cotton growers. Like its frugal and hardy forebear, Georgia Green Seed, it grew as well on red clay soils of the Eastern states as in the rich black loams of Louisiana, Mississippi and Alabama. It produced good crops of commercially valuable lint on all types of soils suitable for raising cotton, and was comparatively immune to the various plant diseases known at the time. Consequently, the new Mississippi variety was introduced into all the cotton growing states in the early 1830s and soon replaced all the older varieties. Thus during that decade a cross between the old Georgia Upland and the imported Mexican cotton became the standard breed of ante-bellum American cotton and parent of most of the modern American strains as well. Because this Mississippi cotton resembled its Mexican ancestor in appearance rather more than it did the Green Seed, it became known popularly if unaccurately as the "common Mexican cotton."

The so-called Mexican cotton of Mississippi was further improved during the 1830s through the work of a group of able planters living in the "Gulf Hills" region of Mississippi not far from the town of Rodney. Seed of this hybrid obtained from the stock of Walter Burling was introduced originally into this part of the state in 1824 by Lewellyn Price. As he had been able to obtain only one small parcel of Burling's Mexican strain, Price was compelled to cultivate these specimens for several years under virtually garden-like conditions before obtaining a supply of seed large enough for extensive planting and experimentation. Price's Mexican cotton, according to the testimony of a near neighbor, improved noticeably under intensive cultivation in very rich freshly cleared soil, greatly increasing in yield and in quality of fiber.[28]

After Price had been planting his Mexican cotton for several years, Dr. Rush Nutt and others obtained some seed from him. Nutt, who was an experimenter of importance, introduced the practice of carefully selecting the cotton seed intended for planting into Mississippi, and his innovation soon became a commonplace among his neighbors of the Gulf Hills. The technique employed by these pioneer seed breeders consisted of choosing for planting purposes only the largest and best formed of fuzzy white seeds; they scrupulously rejected small, immature or misshapen seeds as well as those showing any characteristics of the old Green Seed or Black Seed varieties. By this primitive method of selective breeding Nutt and his friends were able to preserve the best qualities their cotton had inherited from its Mexican ancestor. In this way too they kept it from "degenerating," as they termed the Mexican hybrid's tendency to revert to the properties of its inferior Green Seed parent stock.[29]

Planting seed of the Mexican hybrid variety from the Rodney area went on sale in the year 1833 in market towns as far apart as New Orleans and Augusta, Georgia. After a trial of only one season, the Petit Gulf cotton, as this hybrid became known commercially, quickly won a highly favorable reputation wherever it was planted. The time of its introduction into the Eastern states was exceptionally opportune. In 1834 the common Green Seed Upland cotton of the East became badly infested with the rot, which had moved eastward from the Mississippi Valley; and crop yields of this traditional variety of cotton were reduced in many areas by as much as thirty percent. Those who planted the Petit Gulf in that year for the first time discovered that this Mississippi cotton was not subject to the disease even when growing adjacent to stands of infected Green Seed cotton.[30]

Immunity to the rot alone would have been sufficient to win the hearty approval of Eastern cotton growers for Mississippi Petit Gulf, but they found that it had other desirable qualities as well. On the red clay soils of the East, Petit Gulf produced more cotton of a higher quality than did undamaged Green Seed, and the Southwestern cotton was easier to gather. Progressive Eastern farmers in subsequent years, therefore, began to import this special breed of the famous Mexican cotton from the distributing point at New Orleans, and many of them made a practice of renewing their stock at regular intervals in order to check the Petit Gulf's habit of degenerating after growing for several seasons on inferior soils of the Eastern Piedmont. In consequence the Eastern states soon became an important market for the seed breeders of the Mississippi Gulf Hills, and the trade in planting seed that developed during the closing years of the 1830s made Mississippi the Old South's major breeder and supplier of high grade cottons.

CHAPTER III

The Formative Period of Agriculture

After successfully adopting a commercial type of agriculture based upon cotton in the closing years of the 1790s, the farmers and planters of Mississippi passed through a period of experimentation lasting more than three decades. Beginning in 1800 with farming methods inherited from the tobacco era, the pioneer cotton growers learned in the hard school of practical experience how to cultivate their new crop with greatly enhanced efficiency. Good land was plentiful and cheap in the years leading up to the depression of 1837-49, while labor was scarce and expensive. Planters and farmers inevitably, therefore, put most of their emphasis on increasing their acreages in cotton and corn. They gave little or no attention to developing methods of farming that would increase the yield per acre. From the outset, Mississippi agriculture was like American agriculture in general: it was extensive and exploitive rather than intensive in nature.

The evils always inherent in an extensive system of clean cultivation of staple crops soon became apparent in Mississippi. Both land erosion and soil exhaustion had been characteristic of corn and tobacco culture on the rolling lands of the Natchez District before 1790. When similar methods of farming were applied to cotton, the effect upon the soil was equally destructive.[1] As all three of these crops were kept free from grass by continuous cultivation, the fields were denied the protection from the washing effects of heavy rains which only a well-matted sod can give. Consequently, top soil began to wash away almost as soon as newly cleared fields were put into use. Small gullies became large

37

ravines within a few years, sadly scarring the hillsides.[2] After this ruinous process had progressed far enough to destroy the productivity of a field, the rambling rail fences surrounding it were torn down, and the land was allowed to grow up again in weeds and bushes. New timber or cane land then was cleared and brought under the plow to replace the worn-out field. In retrospect, this unending process of clearing, cultivating, and destroying the fertility of the soil—a process which was one of the principal characteristics of agriculture of the Old South—resembled nothing so much as a cancerous growth spreading death and desolation across the face of the earth. There was unconscious irony indeed in the cotton grower's reference to clearing new ground as "improvement of the land."[3]

Those who follow an agricultural practice which devours the soil from beneath their feet must be ever on the move, and so it was with Mississippi's farmers and planters during the early 1800s. They were agricultural nomads who moved across the land with the ponderous stride of a glacier and, like the glacier, they never stopped advancing. Each winter they cleared new ground in order to expand their operations or to replace land that erosion and exhaustion had made worthless. When their woodlands were gone, they bought land elsewhere and began their vicious cycle again.[4] Compared with the human lifespan, this cycle of agricultural destruction did not take long to run its course. Ten to twenty years of cultivation in row crops usually brought ruin to hill lands in all parts of Mississippi.

Many successful cotton growers of this period were able to anticipate the exhaustion of their soils well in advance. Such planters made a regular practice of investing part of their profits in tracts of land in newly opened areas or in adjoining states. In these cases, slaves were sent out from the headquarters plantation during slack periods of the year to clear land, build cabins, and to make the general preparations necessary for a gradual transfer of farming operations from the old place to the new. In this way cotton growers brought new plantations into cultivation without pressure or haste, while they were squeezing the last vestiges of profit from their capital investments in older tracts.[5] Thus planters with large labor forces frequently worked two or more plantations simultaneously; and it was not uncommon for subsidiary plantations to be owned and operated by partnerships consisting of two or more individuals.[6] In enterprises of the latter type each of the partners ordinarily contributed money and slaves of his own to be used on land owned in common. The management of these plantations was usually entrusted to professional overseers. Regardless of the size or type of operation, there was one common denominator between farms and plantations: their continual cultivation in cotton and corn tended to exhaust the soil.

The transitory nature of agriculture in the ante-bellum South gave little encouragement to construction of permanent buildings, roads or fences. Mississippi farmers or planters seldom built houses in the expectation that they and their heirs would occupy them for generations, as was often the case among the Dutch and German farmers of the Middle states. Apparently the only elaborate and costly houses constructed in Mississippi during the early decades of the nineteenth century were those built by planters of unusual wealth who wanted to establish their residences permanently upon "home places" located conveniently near urban centers like Natchez and Port Gibson. Estates of this kind, with imposing houses and well landscaped grounds, were so few in 1837 that they were actually rare.[7] This disinterest in fine architecture was explained by a Northern visitor in this manner:

> The peculiarity of dwellings of planters . . . may be traced to the original mode of life of most of the occupants, who, though now opulent, have arisen, with but few exceptions, from but comparative obscurity in the world of dollars. Originally occupying log huts in the wilderness, their whole time and attention were engaged in the culture of cotton; and embellishment, either of their cabins or grounds, was wholly disregarded. When they became lords of a domain and a hundred slaves; for many retain their cabins even till then . . . [they] razed the humble cabin, and reared upon its site the walls of an expensive and beautiful fabric. Here the

planter stops. The same cause which originally influ-
enced him to neglect the improvement of his grounds,
still continues to exist. . . .[8]

A vast majority of Mississippi's white inhabitants dwelt in
modest log or frame cottages instead of the classical mansions
of myth and legend.[9] On places belonging to absentee owners
the overseer's cabin was usually the only habitation occupied by
whites, and this dwelling was generally similar in appearance to
the log huts of the nearby "Negro quarter." Even when the
owner himself lived on the premises with his family, more often
than not he occupied a home that was quite unpretentious in
the view of persons familiar with the comfortable farm houses
of the Northern states. The expression, "Mansion," in ante-bellum
Mississippi was used simply to designate the home of a plantation owner. It was synonymous to "Big House," a term sometimes applied to a log cabin which was only a little larger than
the Negro houses around it. Small planters, yeomen farmers, and
cattle ranchers of the pine barrens almost invariably lived in
the typical log cabin of the American frontier everywhere east
of the Mississippi River.[10]

It is significant to note that farm and plantation houses of
ante-bellum Mississippi were not accurate gauges of the wealth
and social position of their owners. As late as 1852, Edward R.
Wells, a Yankee schoolmaster teaching in Vicksburg, noted in
his diary that:

> The Quarters and Mansion houses are like many
> others in this vicinity constructed without reference to
> neatness or architectural effect. But one should not per-
> mit the outward appearance of the buildings which he
> may see during his visits to the plantations of Missis-
> sippi to influence his estimate of the manner of his
> reception or the capability of the planter. The visitor
> will receive princely attention in many of the log-cabins
> of the Southwest, and find intelligence, refinement and
> hospitality, in rooms chinked with mud and scantily
> whitewashed. . . .[11]

Successful Mississippi cotton growers before 1850 usually spent their profits on slaves and land instead of houses and grounds.

Constant preoccupation with cultivating cotton impressed an almost indelible mark upon the minds of Mississippi farmers during the early 1800s. Cotton growers became extremely conservative toward all agricultural innovations. As a rule they were much slower than their Northern contemporaries to take advantage of new scientific discoveries or improved farm implements.[12] This deeply rooted agrarian conservatism was the natural result of employing slave labor in an atmosphere of impermanence engendered by progressive soil erosion. Cotton planters learned from costly experience that expensive or complicated tools, implements, or machines, were likely to suffer damage from the careless handling of unskilled slaves. Nor did the application of scientific farming techniques to extensive cotton farming prove practical. Wherever slave labor was used, these techniques required a higher degree of supervision than Mississippi planters were willing or able to provide. For similar reasons, the care of livestock and the maintenance of plantation roads, fences, buildings, and farming equipment were neglected on most plantations to an extent that would be shocking to farmers of a later age.

Progressive deterioration of soil fertility exerted another curious and lasting influence upon the thinking of Mississippi cotton producers during the years before 1837. Because this evil was a constant factor in their farming experience, they used the ability to wear out farms as a measure for success.[13] With seemingly limitless tracts of fertile land available at low prices elsewhere in Mississippi, Louisiana, Arkansas and the Texas Republic, Mississippi cotton growers had but small incentive to devote time, labor or money to soil conservation. All too frequently, they regarded their soil as a cheap and expendable raw material which, when worked by slaves, could be converted easily into marketable produce. Furthermore, the relatively high price of labor caused farmers and planters to conserve labor and waste land—the cheaper item. Thus early Mississippians were almost unique in history: they were farmers largely devoid of that deep and abiding love for the land characteristic of agricultural peoples everywhere. This want of feeling partly explains why their economic system was ruthlessly exploitive in character during the years leading up to the Civil War.[14]

While few systems of agriculture have been as wasteful of natural resources as that used in Mississippi before 1837, it would be manifestly unfair to hold the cotton growers wholly responsible for the damage which they inflicted upon the land. There was no one to teach them how to farm properly when cultivating staple crops with slave labor. European agricultural experience had been limited almost entirely to growing grains and grasses. Southerners, therefore, could learn little from European farmers about the best methods of raising clean cultured row crops like cotton and corn. Similarly, the farming methods of Northern wheat growers were without useful application in the Lower South where the climate, soil and labor system were radically different from those in the Northern United States. Hence Southern cotton growers were compelled to devise farming techniques of their own which would be suitable to their peculiar commercial agriculture. Knowing no other way to proceed than by the costly method of trial and error, they naturally made numerous mistakes. Yet, despite their many sins of omission and commission, these Southwestern cotton growers somehow managed to achieve a measure of progress in the early decades of the nineteenth century. This progress came mainly from increasing the working efficiency of the labor force through better methods of cultivating cotton, and through developing superior new strains of the staple.

In 1800 no great amount of skill was necessary for growing the hardy and adaptable upland cotton plant. On the other hand, successful cotton raising then, even more than at the present time, required the expenditure of an immense amount of labor. Unlike the Northern staple, wheat, which could be left unattended between planting and harvesting times, the staple crop of the Lower South demanded almost constant attention between April and July. In order to obtain its optimum growth, the young cotton plant had to be free from grass, and the soil around its roots had to be loose. The amount of cultivation needed to achieve these objectives varied with weather conditions during the growing season. Two workings often sufficed in dry seasons. In wet years repeated plowing and hoeing were necessary to prevent the crop from being smothered by grass and weeds. When the cotton plants finally grew large enough to be out of danger from grass, the crop could be "laid by" until the bolls began to open in late August and early September. During these weeks in late summer, the labor forces of farms and plantations were left free to do those tasks which had to be neglected during the weeks of cotton cultivation. In such periods as these, ante-bellum cotton growers harvested their crops of corn, fodder, and peas. As soon as cotton bolls began to open, however, work not of the most urgent character was dropped, and all hands physically able to work in the fields were set to picking cotton. The greater part of the available labor force was employed continually in the fields until all of the crop was gathered, even though the harvesting might sometimes last into the months of January and February of the following year.[15]

At the turn of the century, the usual way of planting cotton was to drop a few seeds into a hill and then cover them with a hoe.[16] Somewhat later, another mode of planting replaced this slow and tedious process in Mississippi. This was a method that Southerners had once used in the growing of indigo and tobacco. As the first step, a shallow furrow was opened with a simple one-horse plow, or "bull tongue," in ground which had been broken up earlier. A sower with a sack of cotton seeds slung over his shoulder followed behind the plow casting handfuls of seeds into the tiny trench. The seeds, on some farms, then were covered over by laborers with hoes. On more progressive places horse-drawn harrows with heavy blocks of wood called "drags" were used instead. By the late 'thirties the efficiency of planting "gangs" of this type had increased to the point where they were able to plant from ten to fifteen acres of land a day without undue effort. On medium sized plantations three or four of these three-man gangs were sufficient to complete the planting of the cotton crop between the first of April and the middle of May.[17]

In Mississippi the growing cotton crop was cultivated almost exclusively with the hoe during the first two decades of the cotton era. In time, however, the plow was brought into general use. As a result the number of workers required to keep the grass down was correspondingly reduced.

A suggestion made by President Thomas Jefferson in regard to plowing methods was of great value to Mississippi cotton producers.[18] While American ambassador to France, Jefferson had been impressed by the merits of plowing horizontally across the face of hills, a system then much in vogue on French farms situated on rolling terrain. Some years after his return to the United States, Jefferson wrote a description of this method of plowing to William Dunbar of Natchez, and expressed his belief that it would be well suited to that part of Mississippi.

Dunbar, like Jefferson, was a child of the Enlightenment. Born in Scotland in 1749, a younger son of a noble family, he emigrated to the New World to seek his fortune after completing an education in mathematics and science at Glasgow and London. The venturesome Scot in 1773 established a large plantation near Baton Rouge in what was then British West Florida, and resided there for many years. Dunbar prospered under many flags, as he knew how to make himself both agreeable and useful to government officials of all nationalities. In 1800 he owned several plantations, and was one of the wealthiest men in the new Territory of Mississippi. Although fully occupied with agricultural affairs, Dunbar remained at heart a scientist. He was a frequent correspondent of scientific academies and an inveterate experimenter. Like Jefferson, he was fond of applying science to agriculture. On his home place, "The Forest," near Natchez, he developed farming implements based upon scientific principles, experimented with new crops and various strains of cotton, and even investigated the possibilities of steam power for plantation machinery. Until his death in 1810, Dunbar was without doubt the principal leader of agricultural reform in the Lower Mississippi Vallay.[19]

The introduction of horizontal plowing into Mississippi was not the least of Dunbar's achievements. In accordance with President Jefferson's suggestion, he performed the experiment of running his rows along the contour lines of hills and found that horizontal plowing was useful in two ways. Level rows reduced erosion. Even more important, horizontalizing permitted plows to be used on fairly steep hillsides.

William Dunbar's experiment in laying out fields with horizontal rows was imitated by cotton growers in many parts of the state during the next quarter century. While there were certain definite advantages to this system of plowing, it was not an altogether effective means of preventing soil erosion. Rows running approximately level were able to retain the water from light showers in their middles, but not the water from heavy rains. Because these furrows had no outlets into a system of drainage ditches, steady downpours quickly filled them to overflowing. When the surplus water ran over the rows at low points, transverse gullies were opened just as destructive as those started by furrows running downhill. Dunbar's system of horizontal plowing, on the other hand, did make possible the use of plows to an extent that had been thought impossible when cotton and corn were planted on steep slopes in irregularly spaced hills or squares. Therefore, adoption of contour plowing was an outward sign that Mississippi farmers were becoming increasingly aware of the value of plows as labor-saving instruments of husbandry.[20]

By the late 1830s plowing in Mississippi had become almost as important as hoeing in the cultivation of cotton and corn. By that time the accepted practice was to plant cotton on beds raised above the ground level for sake of drainage. These beds were constructed by throwing the dirt from several adjacent furrows together with turning plows. Adjoining beds were separated from one another by intervals of four or five feet, a spacing made necessary by the large size of cotton plants. Soon after the seeds sprouted, hoe gangs "thinned" them out to about eighteen inches apart. After this initial working of the crop had been completed, plows were employed to keep the grass down in the wide space between the cotton rows. When grass in the middles had been buried with turning plows, hoe gangs had to "scrape" a space no more than a foot wide near the roots of the young cotton plants. This use of the plow enabled cotton growers to lighten the work of their hoe gangs during the critical period in the growing season, and thus to increase the acreage of cotton land that they could cultivate successfully.[21]

Cultivating additional land in cotton through improved methods would have brought little practical benefit to cotton growers had there not been a corresponding increase in the ability of farm workers to gather the resulting larger crops. Cotton producers, from the very beginning of large-scale cotton planting in Mississippi, had experienced vastly greater difficulty in harvesting their crops than in raising them to maturity. In the early 1800s, for example, Mississippians had been able to produce abundant crops of the fine long staple Black Seed cotton, sometimes attaining yields as high as twenty-two hundred pounds of seed cotton to an acre. Their Black Seed cotton, on the other hand, was so difficult to pick that large portions of their crop often were left ungathered at the close of the harvest season.[22] Under such conditions, increasing the yield without correspondingly improving their capacity to harvest the crop would have done cotton farmers no appreciable good.

The introduction of the Mexican hybrid cotton did much to solve the picking problems of Mississippi producers.[23] The Mexican hybrid variety grew fewer bolls to the plant than did either the Georgia Green Seed or the long staple Creole Black Seed cotton. But Mexican's big bolls produced an unusually large quantity of a lint which was easily detached from the pod. Consequently, the daily picking averages of Mississippi farm workers increased remarkably after the Mexican hybrid was adopted as the standard variety. Near the close of the eighteenth century it was estimated that an average farm worker in the old Natchez District could gather about eight hundred pounds of ginned cotton within a picking season; daily picking averages then were well below seventy-five pounds of seed cotton.[24] By 1837, however, average workers were picking in excess of one hundred and fifty pounds daily, and skilled pickers were gathering three hundred pounds or more.[25] Farmers, who in 1800 had hoped to average two 400-pound bales to each field hand, by 1837 were trying to produce crops of six to eight bales to the hand.[26]

Mississippians made no improvements in their implements during the first third of the century at all comparable to their notable achievements in cotton breeding and cultivation. The hoes and plows of this era were hammered out of soft wrought iron in Ohio and Kentucky foundries and then brought down the Mississippi River, along with shipments of corn, wheat and barrelled salt pork.[27] Southwestern blacksmiths fabricated other implements out of wrought iron sheet or bar stock obtained from the same metal-producing Western states.[28] Dunbar did introduce beautifully made cast iron plows of English manufacture into the Natchez District in 1806.[29] Yet, implements of this type were not widely adopted in Mississippi until the 1840s. Although plows were coming into ever wider use for cultivating growing crops of cotton and corn, few of them were well adapted for that purpose. In most cases they had been designed originally only for opening furrows. Cotton growers of this era seldom tried to replace these crude turning plows with implements built especially for the task of cleaning grass out of growing crops. Implements designed and constructed for this and other specialized purposes did not make an appearance until after 1840.

While the equipment used in cultivating cotton and corn crops underwent few changes before 1837, marked improvements were made in the machines used in Mississippi for preparing cotton for market. The efficiency of cotton gins and presses was vastly enhanced in those years, and great strides were taken in the direction of bringing several machines together under a single roof in combinations which permitted worthwhile saving of labor.

All of the improvements made in plantation ginning equipment before 1837 were due primarily to utilization of better materials and more skillful workmanship, rather than to any extensive modification in Whitney's basic principle. During the 'twenties and 'thirties, efficient running gear, fabricated by newly-developed methods of mass-production in Northern machine shops, replaced the crude hand-made machinery characteristic of Mississippi gins during the early 1800s. The quality of cotton saws (or "rags"), in particular, grew better during these years. The old iron saw blades manufactured by local craftsmen using primitive hand tools were supplanted at this time by cotton saws cut out of cast steel sheet with great precision by steam-powered machine tools. The new mass produced cast steel blades were much cheaper than the iron rags and were much more durable. Moreover, cotton cleaned by gins equipped with steel saws was worth more because they damaged the fiber less than iron saws. On the other hand, comparatively little improvement was made in the effective capacity of gins during the period. In 1837, ordinary ginstands still were cleaning no more than three to five bales a day.[30]

During the 'twenties and 'thirties, planters benefitted from reductions made in the cost of ginning machinery. Factory production brought about a significant decline in the costs of manufacturing gins. These lower costs in turn permitted sizable reductions to be made in the selling price of ginstands and ginning machinery. In 1800 complete gins had cost approximately twelve hundred dollars each in the old Natchez District. In 1837 they cost less than half the 1800 price.[31] At one time in the 1820s, Carver gins, manufactured in Massachusetts, were sold in Mississippi for three hundred and fifty dollars, but this was an exceptionally low price.[32] Ginstands, during the 1830s, ordinarily brought prices somewhere in the vicinity of five hundred dollars each.[33]

While plantation machinery thus was cheaper and more efficient, comparable changes were taking place in the construction of plantation gin houses and in the arrangement of the machinery they contained. Early Mississippi ginstands had been mounted on the ground with only a simple shed overhead to keep off the rain, but plantation gin houses had become comparatively elaborate buildings by 1837. On larger plantations gin houses usually were wooden structures two stories high, with outside dimensions of approximately forty by sixty feet. The ginstand itself occupied part of the second floor and was mounted upon heavy wooden beams built into the floor. The "horse path" supplying motive power for the ginstand was located directly beneath it on the ground level. A room designed for receiving and storing the lint after it had passed through the cleaning process was situated directly behind the ginstand on the upper floor. The large press for baling the cotton was installed on the ground floor beneath the room in which the newly ginned cotton was stored.[34]

Human labor was expensive, and great savings in its use

were achieved by this scheme of installing all the machinery used in preparing cotton for market under a single roof. When presses and ginstands were placed fairly close together, both could be driven by the same power, whether from a horse mill or a steam engine. This efficient arrangement eliminated the need for an additional horse mill required to operate a pair of such machines when installed separately. Locating the cotton press beneath the storeroom (so that the cotton could be thrown directly down into the press box) did away with the problem of transporting the lint from the place of storage to another location to be packed into bales. Under the older method of placing these pieces of machinery in different locations, carrying the loose cotton from one place to another had been one of the cotton planter's most tedious and time consuming tasks. In the new arrangement, no human labor at all was employed in transferring the cotton from the ginstand, for the cleaned cotton was blown directly from the ginstand to the storeroom through a chimney-like flue by means of a strong wind generated by the gin's whirling brushes. While the light cotton fibers were being wafted through the flue, much of the heavier dust and trash that had been mixed with the cotton tended to settle to the bottom of the flue, where it was trapped by grates made of wooden planks. Thus after ginning the cotton fiber passed through a second cleaning process on its way from the gin to the place of storage.[35]

On most of the plantations and farms of pre-war Mississippi horses and mules supplied motive power for gins, presses and the various kinds of mills. During the 'thirties, however, steam engines began to take the place of draft animals in ginhouses on the larger plantations. As early as 1805, William Dunbar had investigated the possibility of attaching a small steam engine to a cotton gin. It was not until 1830, however, that Dr. Rush Nutt, of the "Gulf Hills" region, purchased a steam engine of Pittsburgh manufacture and used it to drive two ginstands on his plantation near Rodney.[36]

Unfortunately little is known about the career of Dr. Rush Nutt, who was an early agricultural reformer second in importance only to William Dunbar. Leaving Virginia in 1805, he first settled in Jefferson County and then moved to the vicinity of Rodney. After his arrival at the "Gulf Hills," Dr. Nutt gave up the practice of medicine and became an outstandingly successful cotton planter. Like Dunbar, Nutt was not bound by convention. He travelled widely in Europe and North Africa, always returning with fresh ideas and samples of seeds collected along the way. He experimented with plants of foreign origin, especially with exotic strains of cotton, and made the product of his plant breeding available at no cost to his neighbors. He seems to have been responsible for introducing the practices of using peas for plant fertilizer and of burying cotton and corn stalks instead of burning them as the old custom had been. Nutt also devoted attention to improving farm implements and machinery, and much of the credit for refining the Whitney cotton gin was given to him by his fellow Mississippians. The best known monument to his original mind, however, is a mansion at Natchez which he designed and his son, Haller Nutt, executed. This building, popularly known as "Nutt's Folly," is octagonal in shape and crowned with a dome of odd design. None of his neighbors seems to have realized that it was patterned after Moslem buildings seen on his trip to Egypt in the early 1830s.[37]

Nutt's successful experiment with steam persuaded other planters who could afford the cost to power their gins with steam engines. By 1837 engines were in operation on several large Mississippi plantations and some of them were driving as many as four ginstands at once. The trend toward wider application of steam power to plantation machinery gained momentum slowly till 1837. Then the scarcity of money brought it to a halt, and it did not get under way again until prosperity returned to the Lower Mississippi Valley in the decade of the 'fifties.[38]

The methods Mississippi cotton producers used to package cotton for shipment underwent frequent changes during the ante-bellum era, and these changes were especially numerous in the period before 1837. During the eighteenth century cotton had been shipped to Europe in long cylindrical canvas bags containing approximately three hundred and fifty pounds of fiber. These containers had been packed in this fashion. The large canvas bag first was dampened with water to stretch it to its maximum capacity. Then the bag was suspended from a raised platform through a round hole cut into the floor. The mouth of the bag was held open and the bag itself was fastened into place by means of a wooden hoop sewn firmly into the canvas. Loose cotton was dumped into the bag and there tamped and trampled by a man working inside. After the container had been filled to the top with cotton, the hoop was removed and the opening of the bag sewn tightly together. When the wet canvas dried, the cotton within was compressed further by shrinkage of the cloth.[39]

This method of packing upland cotton was discarded in the Territory of Mississippi before 1800, although other parts of the South continued to enjoy it long afterward. A new method of packing came into vogue when a practical cotton press was invented by a Natchez mechanic and ginwright named David Greenleaf. The Greenleaf cotton press, which came into general use around the turn of the century, compressed ginned cotton into a rectangular-shaped bale by forcing a block press into a wooden box in the manner of a piston working within a cylinder. The flat-faced follower was driven into the cotton-filled box by two large wooden screws operating in a horizontal plane. Each of the two screws was turned by crews of two or more men using levers thrust through the shank of the screw. While the cotton box was still empty, its inside walls were lined with hempen cloth, which became an outside cover for the finished cotton bale. When the mass of cotton in the press had been squeezed into the desired size and shape, two removable sides of the box were taken off exposing the unfinished bale on two sides. The hempen cloth in the box was drawn tightly around the bale at this point and sewn firmly together. To complete the baling operation, a number of hempen ropes were drawn through grooves cut into the faces of the press box and then tied around the bale in order to hold it in shape. Then the follower was withdrawn from the box, and the completed bale was extracted from the press. When the external pressure on the bale was relieved, the cotton within the bale expanded with enough force to stretch the ropes slightly, giving the cotton bale a shape that was more nearly cylindrical than rectangular. Operation by hand was a laborious task, and only a few bales could be processed in the space of a day with the Greenleaf horizontal screw press.[40] Yet the Greenleaf principle was used to advantage in later improved presses.

Becoming dissatisfied with the performance of the Greenleaf press, William Dunbar drew up specifications for a much more powerful machine, and arranged to have it manufactured by a Philadelphia firm.[41] The finished press, which employed a cast iron screw, was delivered in 1802, and Dunbar erected it on one of his plantations near Natchez. This press with working parts of cast iron had been expensive to make, and Dunbar paid for it by shipping seven bales of cotton, worth one thousand dollars, to the manufacturer in Philadelphia.[42] Dunbar's original purpose in obtaining such a press had been to compress his cotton bales into such a small size as to render further compressing at the seaport completely unnecessary.[43] At the time he also hoped to reimburse himself for the large sum invested in the cast iron press by using it to crush cotton seed and selling the vegetable oil extracted. Little is known, however, about the success of either of Dunbar's path-breaking experiments with this press.[44] Yet, it is known that the practice of pressing oil from cotton seed was not undertaken on a commercial scale in the Natchez area until the 1830s when a company was organized for that purpose. Nor did Dunbar's cast iron press become popular with Mississippi cotton growers, probably because of the high price and the difficulty encountered in having others like it manufactured locally. Cast iron presses did not come into common use until after 1840, and at no time before 1860 were they used as widely in the Old Southwest as the cheaper wooden machines.

Wooden cotton presses multiplied in number and type between 1800 and 1837 as local craftsmen throughout the state directed their attention to solving the cotton growers' baling problems. Several different varieties of these locally manufactured machines were in wide use during this period. One of these, a horizontal cotton press, operated by two heavy wooden screws somewhat like the old Greenleaf press, was popular on large plantations where both presses and ginstands were installed within a single building and could therefore be driven by a common source of power. Presses of this horizontal type possessed the advantage over other similar machines of requiring little room overhead. For this reason they were well adapted to the limited space available in ordinary two-story plantation ginhouses.[45]

A vertical press worked by a single wooden screw was common in sections of the state where farms and small plantations were the rule. Crudely made and clumsy, these machines towered high into the air without any kind of shed to protect them from the weather. The single wooden screw moved the follower in a vertical plane, pressing on the upward or on the downward stroke, depending upon the manner in which the machine was made. The operating screw was turned by means of two long levers fastened to the top of the press and extending downward at an angle in the fashion of the arms of a capital letter "A." Teams of horses or mules were hitched directly to these levers in an arrangement that eliminated the need for any additional operating machinery. Being exposed to wind and rain, vertical presses could be used for baling cotton only on clear days, and much tedious labor was involved in carrying cotton from the ginstand to the press and in hauling it up a ladder to the elevated press box. Carrying the loose fiber from one place to another on clear fall and winter days exposed it almost constantly to the danger of being blown away by the wind.[46]

A third type of baling machine was used occasionally on cotton plantations and farms. This was a form of horizontal press operated by long beam-like levers instead of by the more conventional screws. The power lever was pivoted at one end. This lever drove an attached piston shaft into the press box when it was pulled forward through a great arc by a team of horses or mules. Numerous variations of this principle were used, all of them intended to reduce the great length of the operating lever.[47]

All types of presses used in Mississippi during the 1830s had at least two features in common. All were operated by draft animals or steam engines, and all were capable of packing cotton into comparatively heavy bales. In 1836, for example, the bales pressed in the South Atlantic states as a rule weighed between 300 and 325 pounds; those pressed in Mississippi were heavier, weighing 400 to 500 pounds, or even more. As transportation charges in the ante-bellum period were levied against the number of bales rather than against the total weight of the cotton shipped, the greater weight of Mississippi cotton bales resulted in worthwhile savings in shipping costs.[43]

After the Mississippi cotton crop of the pre-depression era had been gathered, ginned and packed into bales, it was carried by water to the great Southwestern cotton market at New Orleans. On this difficult first leg of the long journey from Mississippi farms to textile manufacturing centers abroad, cotton bales were hauled overland to the water shipping point in clumsy, slow-moving wagons drawn by teams of twelve to sixteen oxen. Because dry roads and favorable weather could be counted on only in the early weeks of fall, planters and farmers made a practice of beginning to haul their cotton to the river bank as soon as the first few bales of the crop were ready for shipment. When October lengthened into November and the fall rains commenced, Mississippi's roads rapidly turned into quagmires. As winter came on and great numbers of heavy cotton wagons wound their way to the river landing stages, the ruts in the roads grew deeper and the quagmires often became nearly impassable. Because of bad roads the task of transporting cotton across country was one of the most difficult with which cotton growers had to contend. Men, wagons and teams were sometimes strained almost to the breaking point by the cold rains and mud of Mississippi winters.[49]

After delivery to the river bank, the cotton bales were loaded on a river boat for the second stage of the journey. Until about 1820, cotton was carried from the old Natchez District down the Mississippi River in unpowered keel boats and flat boats of the type used to bring produce from the Ohio to markets along the lower reaches of the Mississippi.[50] In the 1820s and 1830s these cotton carriers were replaced by steamboats on the

Mississippi River and its larger tributaries, as well as on the Pearl and Tombigbee Rivers to the east.[51] The old-fashioned unpowered keel boats nevertheless continued to perform useful services; for they were the only craft that could reach Mississippians living above the head of steamboat navigation, or along creeks too narrow to permit the passage of larger river vessels. When these small streams were filled by the heavy rains of winter and early spring, keel boats laden with merchandise were brought upstream far above the points served by steamboat transportation. After discharging their cargoes, these odd craft took on cotton which had been deposited on the banks of the stream earlier in the year. Then the heavily laden boats drifted downstream to the head of steam navigation, where the cotton was transferred to steamboats for the final leg of the journey to New Orleans. Long, narrow river boats of this type remained the only means available to many inland inhabitants for transporting bulky merchandise until Mississippi's railroads were built.[52]

After delivery to warehouses in New Orleans, the cotton was sold either to agents of English textile manufacturers or to professional cotton buyers who made their living by purchasing cotton in New Orleans for resale in the markets of Europe. From New Orleans the cotton was then shipped to the manufacturing centers of New England, Great Britain and the continent. In the 'twenties and 'thirties, cotton consigned to overseas destinations was re-baled in New Orleans before being placed on transoceanic sailing ships. This second compressing was performed with powerful steam driven machinery, and resulted in reducing the bale to approximately half its original dimensions. These recompressed bales were superior to the bulkier packages turned out by plantation presses, for they were less likely to suffer damage when exposed to fire or water and they occupied less storage space.[53]

In the 1830s steam powered cotton compresses, like the ones in New Orleans, were erected in the Mississippi towns of Natchez and Grand Gulf; and ocean-going sailing vessels engaged in the cotton carrying trade were sometimes towed up the Mississippi River by small steam tug boats to loading points as far inland as Vicksburg. These facilities made it possible for a small portion of the Mississippi cotton crop to be shipped directly to European destinations without making the usual stopover at New Orleans. Yet the direct cotton trade carried on from Natchez, Grand Gulf and Vicksburg to Liverpool never attained a volume sufficient to threaten the supremacy of New Orleans as the cotton exporting capital of the Old Southwest. Throughout the ante-bellum period, by far the greater portion of Mississippi's cotton as well as the crops raised in Louisiana, Arkansas and Tennessee were consigned to the Crescent City for sale and transshipment to markets both at home and abroad.[54]

Cheap and convenient water transportation exerted a profound influence upon the economic development of Mississippi during most of the ante-bellum period. The Mississippi and Yazoo rivers, with their tributaries, provided the western and northern parts of the state with an extensive system of navigable waterways reaching far into the interior. In the southern and eastern portions of Mississippi, the Pearl and Tombigbee rivers opened water highways to the Gulf of Mexico. In fact, few regions in the state suitable for commercial agriculture were located beyond practical hauling distance from some navigable waterway.

Mississippi's extensive system of creeks and rivers served the cotton growers of the state in other ways than transporting their staple crop to market. The waterways brought the livestock, grains, metals and manufactured products of the Old Northwest to them; and made these commodities available at prices which bore reasonable relation to the generally prevailing prices of cotton before 1837. As early as 1790, cheap produce from Kentucky, introduced by General James Wilkinson, had driven from the New Orleans market the corn, pork, beef and tobacco produced in the old Natchez District. A similar price differential on foodstuffs continued to exist as late as 1837. Mississippi farmers and planters, as a consequence, were inclined to devote most of their available supply of labor to the task of raising cotton. Because they could get foodstuffs from the Upper Mississippi Valley at moderate expense, they made no consistent or determined effort to raise their own grain and meat. With a large number of farmers pursuing this general policy, Mississippi, down to 1837, remained a large importer of livestock of all kinds, salted beef and pork, corn, wheat, hay, bar and sheet iron, axes, hoes, spades, nails, horseshoes, plows, rope, hempen cloth, coarse cotton and woolen fabrics for Negro clothing, and even Negro slaves. To pay for this imposing list of articles, Mississippi obtained cash in quantity from only one source—the sale of raw cotton.[55]

Although Mississippians were constant purchasers of Western produce before 1837, they did not wholly neglect to raise food on their farms and plantations. The truth, in fact, was quite the reverse. Considerable quantities of an upland variety of rice were raised for home consumption in the southern part of the state throughout this period, and wheat was grown in the northern hill counties after their settlement began in earnest during the 1830s.[56] Sweet potatoes were produced in large quantities on all farms and plantations in the 'twenties and 'thirties, and cow peas, or Indian peas, as this plant sometimes was called, were rapidly becoming a major local source for livestock feed. Corn, however, was by far the most important food crop for man or beast. It was grown in all parts of the state during the entire ante-bellum period, on agricultural establishments of all sizes and types, from the simple clearing of the hunter to the sprawling plantation of the great slaveowner, in quantities and importance second only to cotton.[57] In the years before the Civil War, Mississippians cultivated corn in much the same manner as cotton.[58] They planted it on raised beds, or rows, in drills; and worked it with the hoe and the plow in order to keep down the grass between the rows. Corn was a grain which reached maturity rapidly, and two crops could be harvested in a single season in latitudes as far north as Mississippi. Furthermore, corn yields were remarkably abundant during the early years of the cotton era, comparing well with crops of the present time. On the new fields of rich river bottom lands, for example, corn in the first season or two usually produced as much as sixty or seventy bushels to the acre. Corn production declined as the fertility of the land was reduced progressively by erosion and exploitive methods of cultivation. Yields of thirty bushels to the acre were considered normal on "improved" hill lands that had been in cultivation a few years. Cropped-out fields in the hills, after having been under the plow continuously for ten to twenty years without manuring of any kind, could be expected to produce as little as ten bushels to the acre, or even less.[59]

The early crop of corn was planted in March or April, in order to reach maturity during the slack period in the cotton season, which came in August. The late crop, planted in May or June, was left standing in the fields through the fall, and was not gathered until most of the cotton had been picked in December and January. Many cotton growers of the time followed a queer practice in regard to the fall crop. When the corn was fully matured, farmers bent the ripe ears upside down upon the stalk. They believed that, with the ear thus inverted, the shuck would shed rain water more effectively and protect the grain from the attacks of birds and insect pests. By this practice, the harvesting of corn could be delayed until the cotton was gathered.

Vast quantities of green corn leaves from the growing plant, which were referred to as "blades," were also harvested in late July or early August, then dried or cured, and fed to livestock in much the same fashion as hay in the Northern states. Dried corn leaves were fed to farm animals, along with ear corn, during the working season when there was no time for grazing, and also during the late fall, winter, and early spring, when there was insufficient grass to sustain them in good condition. Thoughtful farmers suspected that pulling fodder from the growing plant was injurious to the grain, and a few of them followed the practice of sowing corn broadcast to be cut green and used as hay. A majority of Mississippi farmers, however, took the risk of lowering their grain yield by pulling fodder in the traditional manner.[60]

While corn remained the largest food crop and the second largest crop of any kind produced in Mississippi throughout the ante-bellum period, there was a relative decline in its cultivation during the boom period of the early 'thirties. The trend at that time, at least on large cotton plantations with ready access to water transportation facilities, was in the direction of increasing the importation of corn from the Upper Mississippi Valley, thus providing additional acreage for cotton, which was bringing extraordinarily high prices. During those years, the farmers of at least one inland county—Hinds—chose to haul their corn from the river port at Vicksburg over many miles of incredibly bad roads in preference to reducing their cotton acreage to raise their own food supplies.[61]

Two major agricultural innovations were introduced into Mississippi during the late 'twenties and early 'thirties, and both were applied first to the cultivation of corn. Both practices were aimed at improving the fertility of the soil, or slowing down its rate of declining productivity. The first innovation consisted of using cotton seeds as fertilizer. During the first few decades of the cotton era, most Mississippi farmers had considered cotton seed a nuisance. But in time Mississippians learned that they were extremely valuable as a fertilizer for growing crops. Just when this important discovery was made, and by whom, is unknown. By the early 'thirties, however, cotton growers were increasing their yields of corn by burying cotton seed near its roots. The practice spread from the Rodney area, where it seems to have originated, to other parts of the state during the next few years, and it was applied to crops other than corn. By the 1850s, cotton seed was regarded as the universal fertilizer in Mississippi.[62]

A second innovation of almost equal importance to the first was made by Dr. Rush Nutt. In the early 1820s Nutt planted peas between rows of growing corn.[63] Planted during the last working given to the corn crop before it was "laid by" for the year, the peas made their early growth while the corn was ripening and did not attain full growth until after the latter had been gathered. When fully matured, the peavines covered the ground between the rows and entwined themselves around the standing stalks of corn. After the corn had been harvested and the peas had ripened, the livestock was turned into the field to graze upon the luxuriant vines. Horses, mules and cattle ate the peavines greedily, and hogs throve on the ripened peas. Cowpea vines grown in a corn field after this fashion were useful to the farmer in a number of ways: in addition to providing food for stock at a time when natural pastures were beginning to fail, the peavine gave valuable protection to the soil by shading it from the hot August sun and by reducing the extent of erosion caused by late summer showers. Even more important, the cowpea improved the fertility of the land by extracting nitrogen from the air and adding it to the composition of the soil.[64]

Although farmers did not learn the secret of the nitrogen-fixing quality of the legume until the closing years of the ante-bellum era, they nevertheless discovered at an early date that fields would become more productive in cotton after being planted in corn and cowpeas. Perhaps not unnaturally they made the error of attributing the improvement in soil fertility to the corn, instead of the cowpeas. As a result, many recommended that land which had been planted in cotton for several years be cultivated for one season at least in corn. The prevailing custom of following corn with cowpeas in the manner described above caused the cotton-corn theory of rotation to seem to possess some validity.

By the mid 'forties the ability of the cowpea to enrich the soil had become so well known to Southern farmers that they gave it the title "clover of the South." Yet statistics of the period do not reveal to what extent this legume actually was cultivated in Mississippi. Census figures list only the harvested peas, thus ignoring a large part of the crop which was consumed in the field by the stock. As ante-bellum farmers were in the habit of using the cowpea almost entirely in the latter way, very few of the ripe peas were harvested, probably only enough to supply planting seed for the next season. It is certain, however, that the cowpea crop was large.

Probably ranking in importance just above the cowpeas among agricultural productions of Mississippi was the sweet potato, figures on the production of which are also lacking for the period ending with the panic of 1837.[65]

The highly nutritious sweet potato, corn, and salt pork were the basic elements in the diet of a vast majority of the inhabitants of Mississippi at this time, both free persons and slaves.

In that era vegetables, fruits, fresh meat, fish and game appeared on tables when available, and white bread from wheat flour was highly prized when obtainable. But corn, sweet potatoes and salt pork provided the principal items of food. Significantly, the cultivation of the sweet potato was ideally suited to Mississippi cotton farms and plantations. It required very little labor and could be left unattended between planting and harvesting, and the yield was so high that only a few acres were needed to supply a large labor force. It was mainly for these reasons that the plant was grown universally on Mississippi farms during the period before the Civil War.[66]

No crops other than cotton, corn, cowpeas, sweet potatoes, wheat and rice were raised in Mississippi to any great extent before 1837. Tobacco, pasture grasses like Bermuda and timothy, and small grains like oats, rye, buckwheat and barley were known to the Southwestern farming community generally, but they did not become popular with Mississippians during this period. While vegetables and fruits of many varieties were produced in gardens and small orchards, they were not cultivated on a commercial scale. Truck gardening and commercial fruit raising did not become part of Mississippi's economic structure until after the panic of 1837 had made cotton planting less profitable, and even then on a limited scale. Probably at no time before the Civil War were fruits and vegetables grown in Mississippi in quantities sufficient to provide the population with a balanced diet.

Although Mississippians were steady consumers of livestock, dairy products and cured meats produced in the Upper Mississippi Valley, many of them raised livestock in numbers that were by no means negligible. Scrub cattle, horses, sheep and swine descended from stock imported originally from Spain, ranged at large in great numbers along the banks of streams and in wooded areas. Large plantations in the old Natchez District, for example, often possessed as many as five hundred head of the brown native cattle. These beasts were called by the name of the Attapakas or Opelousas districts of Southeastern Louisiana, where they supposedly originated. Grass-fed animals of this breed produced an inferior grade of beef, but they were able to withstand the effects of the hot Mississippi climate and could subsist the year round on native grass and cane with little cost to the owner. Opelousas cattle made hardy and powerful work oxen which were much esteemed by farmers. Purebred imported stock of the kind then becoming famous in Kentucky were unknown in Mississippi, except among a very few planters who had become interested in the breeding of improved livestock.[67]

Indeed, cattle raising was the principal industry in Southeastern Mississippi for many years. This comparatively unknown and unsettled part of the state was a region of pine forests and sandy soil totally unsuited to the growing of cotton and corn except for occasional fertile alluvial soils along the banks of rivers. A profusion of wild grasses covered the floors of the great pine forests, and there were vast canebreaks in hollows and along the many streams flowing through the region. Here cattle had roamed the woods and grazed upon the open range since the latter days of the French occupation of the Mississippi Gulf Coast. When the Indians were driven from this part of the state in the 1790s and the early years of the nineteenth century, they were followed by whites whose occupation was raising cattle for export to overseas markets.

This early Mississippi cattle industry was similar in many respects to one that flourished at a later date on the Western plains. Herds numbering as many as a thousand head grazed at will the year round on the unfenced meadows, swamps and canebreaks that reached inland from the coast for a distance of a hundred miles through the pine barrens. Stock required no feed in this subtropical climate, but blocks of salt were placed at selected locations on the range and in most areas provided the only signs of domestication. Lariat-and-whip-swinging cowboys mounted on wiry ponies gathered the herds in periodic roundups, and branded and tallied the cattle in typical cowboy fashion. On occasions, marketable animals were separated from the herd and driven overland for sale in the great Southwestern cattle market at Mobile. Here the animals were slaughtered, and the meat, hides and tallow shipped by water to New Orleans and the West Indies.[68]

Mississippi's cattle kingdom, like its later Western counterpart, had a comparatively short life. Because of damage to the open range caused by brush and grass fires and by over-grazing, the once great herds of cattle dwindled in numbers during the 1840s and had practically disappeared by 1860. Meanwhile, the people of the pine barrens region had been turning to new pursuits based on lumber, tar, and turpentine. In the 1840s and 1850s the pine forests supplied every year greater and greater quantities of naval stores and pine lumber.[69]

Mississippi sheep of this period were a scrawny but hardy breed of animals that could subsist on the open range without winter feeding. Their wool was so scanty and so coarse as to be almost useless for manufacturing cloth, but their mutton was considered by connoisseurs to be of an unusually fine flavor. Hence small flocks were kept as a mutton supply on many farms and plantations in the older parts of the state, and were being introduced into the northern region in limited numbers toward the close of the period. Only two flocks of fine Saxony or Merino sheep are known to have existed in Mississippi before 1837. One of these was maintained on a place in Madison County belonging to Mark R. Cockrill, who later became one of the most famous livestock breeders of the Old South.[70] The other flock was the property of Benjamin L. C. Wailes, who lived near Natchez.[71] In neither case, however, did the example of these two stockmen find much favor with their neighbors.

The hogs that could be seen rooting in the corn fields and in thickets of oak trees all over Mississippi during the pre-depression era were of the variety that since has become famous as "Arkansas Razorbacks," a name that was self-explanatory. Often these swine were referred to humorously as "Land Pikes," because of their alleged resemblance to the ugly long-snouted fish of the same name. Fully grown specimens of this semi-wild species were small and seldom attained a weight greater than one hundred and fifty pounds. They could be raised in large numbers, however, as their maintenance cost nothing. As a rule, "Land Pikes" were not butchered until they were three years old. During most of their lifetime they were allowed to roam the open fields and woodlands the year round. Grass, roots, berries, nuts, acorns, field peas and cotton seeds provided their food on the open range. Young males not intended for breeding purposes were castrated in order to prevent their becoming ill-tempered blood brothers to the dangerous wild boars of the hills and swamps. When animals for butchering became three years old, they were rounded up, penned, and fed upon ear corn, kitchen swill, and such other foods as the places could provide until they were fat enough for slaughtering. Most of the pork killed in Mississippi before 1837 was not salted but eaten when the meat was still fresh. Farmers and planters of the Natchez area had discovered quite early that the winters of that latitude were not cold enough to permit proper curing of smoked and salted beef and pork, and, consequently, had adopted the practice of importing salt pork from colder states of the North. Planters of Mississippi did not begin to preserve their own pork in large quantities until the depression years of 1837–49. At no time before 1860 were they able to salt enough to meet the state's minimum food requirements.[72]

The small native breed of horses raised in South Mississippi, called "cane tackeys," were degenerate descendants of animals brought into New Spain by the Conquistadores. They were very like their kinsmen, the mustangs of the Texas plains. Tackeys were used mainly by cattle herdsmen of the pine barrens, and were employed infrequently by farmers. Horses and mules used for pulling the plow in that state were almost invariably imported from the stock breeding sections of Tennessee and Kentucky. Fine thoroughbred saddle and race horses were bred occasionally by wealthy Natchez planter-sportsmen, but work horses seldom were raised in Mississippi, and mules not at all. Indeed, draft animals other than oxen were not produced in Mississippi in significant numbers during the pre-depression period.[73]

When comparisons are attempted between agricultural practices on small farms and those on large plantations, few fundamental differences can be found. The primary purpose of both farmer and planter was to grow as much cotton for the market as possible with the labor force available. Both groups supplied their laborers and livestock with corn and sweet potatoes raised in part on the premises. Both employed the system of clean cultured rows in raising their major crops, and both used plows and hoes as principal instruments of husbandry. As they both employed the same system of cultivating crops, they both suffered alike from the evils of erosion and crop exhaustion, and the same livestock—horses, mules, cattle, hogs, and sometimes sheep—were to be found on a majority of Mississippi's farms and plantations of all sizes. And farm hands, whether employed on plantations or farms, could cultivate approximately the same acreages in corn and cotton—seldom more than twenty acres to the laborer.

The small cotton farms so numerous in Mississippi in the 1830s should not be confused with Northern establishments of comparable size. The small Mississippi farm was essentially a cotton plantation, lacking only the Negro slaves. The system of balancing grain production with livestock raising, which was typical of the more advanced types of Northern farms, was not characteristic of Mississippi farms worked principally by white labor. Nor did Mississippi small farmers cherish any ideals about soil conservation and diversified agriculture. Instead, their operations were patterned, so far as their means permitted, upon the money-making commercial cotton plantation.[74]

The most striking difference between small farms and large plantations in pre-depression Mississippi, aside from the obvious matter of size, was in the organization and composition of their labor forces. On units large enough to be designated correctly as cotton plantations, there were always enough Negro slaves to require a special system of social and labor control. On establishments of yeoman farmers, on the other hand, most of the work was done by white members of the owner's family. On such farms, few slaves were employed; and, when they were, they worked alongside the white farm laborers without any special supervision. In brief, the distinguishing characteristic of a true cotton farm was the employment of free white labor—with or without additional help from an occasional Negro slave. On true cotton plantations, however, no white labor was employed except highly skilled workmen and overseers, or perhaps an occasional person in a special capacity.

Mississippi cotton planters of the 1800-37 period had inherited a fully developed system of farm administration and labor control from tobacco growers of the Spanish era, and they had derived their basic farming techniques and crops from the same source. They found this agricultural system to be just as well adapted to commercial cotton growing as it had been earlier to extensive tobacco cultivation. Consequently, the plantation system was employed in Mississippi until the Civil War without any fundamental change.

In many respects a large plantation was an industrial type of operation, functioning within an agricultural setting; and management was provided by the plantation owner, usually with the aid of hired professional white overseers. As in all enterprises of the industrial variety, there was a certain degree of specialization of labor. Routine tasks like wood-cutting, rail-splitting, fence building, clearing of land, planting, cultivating and harvesting, were performed by Negroes who had been trained for those particular jobs. Specialists known as "axe hands," "hoe hands," and "plow hands" made up the bulk of the plantation's force of "field hands," so designated to distinguish ordinary plantation laborers from such skilled Negro workmen as ginwrights, blacksmiths, carpenters, hostlers, cooks, tailors, seamstresses, weavers, shoemakers, and household servants. During the fall almost all of these slaves except house servants were sent to the fields to pick cotton, the most important task on the plantation. At other seasons of the year, the various tasks involving the large scale raising of cotton and corn were performed by "gangs" of trained "hands" working under the direction and control of a "driver," a Negro slave with duties similar to those of a foreman in a factory.

During this formative period of Mississippi agriculture, the white overseer was an especially important cog in the plantation machine. His position combined the duties of a prison warden with those of a farm manager, and it was his lot to supervise and regulate almost every aspect of the lives of the slaves committed to his charge. In addition, he had to make certain that they performed the manifold chores necessary for producing a satisfactory crop of cotton and corn. Although overseers performed vital services in the cotton plantation system, their positions were rewarding neither in a social nor in a financial sense. Even when charged—as they often were—with the responsibility for property and slaves worth more than one hundred thousand dollars, their yearly salary seldom exceeded five hundred dollars, plus room and board. Nor was their social position commensurate with their heavy responsibilities. Most of them were required by their employers to be present on the plantations at all times, and they were expected to be on duty all around the clock. Because it was underpaid and overburdened the occupation of "overseeing" failed to attract enough able and ambitious men of the caliber that large scale enterprises require. Consequently, the efficiency of large plantations was often impaired by poor management.[76]

Yet there were numerous instances in which superior overseers were able to make the transition from employee to slaveholding landowner, and at least a few of these became wealthy planters during the prosperous years before 1837. Thomas Hall of Adams County was an outstanding example. Soon after his arrival in the Mississippi Territory in 1806, he obtained a position as overseer on a plantation belonging to the prominent Minor family. Unlike the ordinary run of overseers who retained a job for only one or two seasons, Hall became an almost permanent fixture on the Minor lands. After serving as an overseer for eighteen years, he left the employ of this family and became a slave-holding plantation owner himself. In the 'twenties and 'thirties he prospered and became well known and respected in the Natchez area, and was one of the leaders in introducing agricultural improvements into his neighborhood. He was perhaps the first to import thoroughbred cattle and hogs from England into Mississippi. After the years of depression, his success continued and, by 1852, his annual crop of cotton had grown to eight hundred bales.[77]

Although the road to wealth in Mississippi was not easy to travel during the 1820s and 1830s, it was nevertheless open to ambitious small farmers and planters. Many who were thrifty enough to accumulate small amounts of capital earned repute as successful cotton growers.[78] During these years, good land was plentiful and relatively cheap, cotton prices were extraordinarily favorable to growers, and credit was fairly easy for those who wanted to buy lands and Negroes. The usual term for loans required repayment within four years; but nobody seems to have considered such short terms unreasonable. Two or three good cotton crops in succession would clear away the debts on a new plantation and bring the means of acquiring a fortune within the landowner's reach.[79]

Favorable economic conditions in Mississippi during the 'twenties and 'thirties attracted settlers in growing numbers from the North and East; and the rising tide of immigration swelled to a flood in 1835, 1836 and 1837. In those years cotton prices reached an all-time peak for the ante-bellum period, and the lands acquired by the removal of the Choctaw and Chickasaw Indian tribes were being thrown onto the open market. Land speculators and prospective farmers and planters hastened to the "New Purchase" in the northern half of the state seeking to acquire choice sites for farms and plantations.[80] Slaves in large numbers were brought into this area from the older states to the east and from the southern part of Mississippi. Towns sprang up in the wilderness overnight and then disappeared just as quickly when the flood of land hunters passed them by. Land prices sky-rocketed, slave prices soared, food and other necessities became very expensive. Tens of thousands of acres of new ground were hurriedly cleared and brought under cultivation in cotton and corn. Fortunes were made and lost with astonishing rapidity through speculation in land. Indeed, conditions were much like those in California during the gold rush, except that the scale was much smaller in Mississippi. But, like many other periods of frenzied prosperity, the great Mississippi boom of 1835-37 was followed by a devastating panic that was just as violent as the "flush times" preceding it.[81]

The panic of 1837 ended the second phase of ante-bellum Mississippi agriculture.

CHAPTER IV

The 1837-1849 Depression and the Agricultural Reform Movement

The financial panic of 1837 terminated an era of almost uninterrupted prosperity and agricultural expansion in Mississippi. Except for brief intervals during the Napoleonic wars and in the short-lived world-wide depression that had followed directly afterwards, prices for Southwestern cotton had remained high.[1] The population of the state had expanded rapidly as a result of favorable cotton prices and an abundant supply of soil suitable for growing the staple. Between 1810 and 1830 the number of white residents had risen from 38,925 to 70,443, and the number of slaves from 17,371 to 65,659.[2]

By 1833 the moderate prosperity of the 1820s had begun to assume the proportions of a boom. After hovering around the ten cent level for most of a decade, the price of New Orleans cotton began to climb rapidly in the fall of 1833, reaching sixteen cents a pound in 1834 and twenty cents in 1835-36.[3] Settlers, eager to take advantage of the profits that were assured when cotton was selling above fifteen cents, streamed into the recently opened northern half of the state in large numbers, bringing with them hundreds of Negro slaves.[4] By 1836 the population had leaped to 144,351 white persons and 164,393 slaves; and in that year Mississippians produced 317,793 bales of cotton (400 pounds each) on 1,048,530 acres of land.[5]

The boom collapsed in 1837. In 1836 the federal government had adopted a new deflationary monetary policy, which was designed to eliminate the easy credit that had been financing the land rush into the New Purchase region in North Mississippi recently ceded by the Choctaws and Chickasaws. Almost at once the price of cotton faltered, and then skidded downward from twenty-one to thirteen cents a pound.[6] Although 66,906 more inhabitants, black and white, were added to Mississippi's population between 1836 and 1840, it appears likely that a majority of the newcomers arrived in the state during the year of the crash, rather than afterward;[7] for the panic not only curtailed the rate of Mississippi's internal expansion, but also reduced the flow of immigration.[8]

When the hard money policy announced by the federal government in 1836 went into effect the following year, it immediately exerted a powerful influence upon the economic development of the Old Southwest. President Andrew Jackson's "specie circular" forbidding the Treasury to receive payment for government lands in currency other than gold coin put an end to a frenzy of speculation in land that had been raging in Mississippi for half a decade. Until then speculation in land had been considered a nearly foolproof means of acquiring wealth.[9] In the days of the land rush into Northern Mississippi, persons with capital or credit had been able to secure vast tracts of land in the Choctaw and Chickasaw cessions by making a relatively small down payment (usually one-fourth of the price) at the time of the purchase; and the remainder was usually paid in three annual installments. The speculators then sold the lands to a host of would-be cotton farmers who were streaming into the New Purchase from all the states of the Old South; and they used their profits to meet the remaining payments as they fell due.[10]

After the "specie circular" was issued, land speculation in the Choctaw and Chickasaw Purchases ceased to be profitable. Neither the land hungry newcomers, nor the speculators, nor the state banks themselves were able to obtain enough specie to comply with the terms of the government's new land policy. Although paper notes issued by local banks circulated abundantly in Mississippi, gold and silver coin was scarce. Consequently, most land speculators were unable to pay specie to the federal land office as required, and were compelled to forfeit title to tens of thousands of acres of choice lands in the Indian cession. Although the total acreage forfeited is unknown, the amount that reverted to the government was certainly very large. Notices of government sales of reclaimed lands occupied so much space in Mississippi newspapers as to give them the appearance of federal land office bulletins.[11]

Although the net effect of President Jackson's policy was to discourage immigration into the Old Southwest, it nevertheless performed at least one beneficial service for the agricultural community of Mississippi. It served to eliminate the ubiquitous land speculator from the region. Thereafter, farmers and planters were able to purchase land more cheaply from the federal government than they had ever been able to buy it from speculators. Some *bona fide* Mississippi cotton growers, on the other hand, were affected just as adversely by the federal government's deflationary measure as the speculators themselves. Farmers and planters, who had bought lands directly from the government prior to 1837 and had not completed their payments before the issuance of the circular, were placed in an impossible financial difficulty when they had to pay specie; and many of them lost their farms.

Sad though the contraction of the money supply and the drying up of bank credit made the plight of the cotton grower in 1837, falling prices during the next decade were to make it even worse. Over-expansion by English textile manufacturers in the 1830s had resulted in flooding the world market with cotton goods, forcing many of the mills to close. This depression in the textile trade played havoc with American cotton prices. They began to decline in 1837 and, for the next eight years, continued an erratic but generally downward course. From a high point of twenty-one cents a pound in 1836, prices for New Orleans cotton dropped to thirteen cents in the fall of 1837. They rose again to fourteen cents in 1838 and to eighteen cents in 1839. In 1840, however, prices began to slide downward again, and this trend continued until bottom was reached in 1845. New Orleans cotton brought from ten to eleven cents between 1840 and 1843, seven to eight cents in 1844, and only five cents a pound in 1845.

Recovery finally commenced in 1846; that year prices rose to eight cents a pound. In the next year they reached the twelve cents level, only to drop back to eight cents in 1848 and 1849. Prices of New Orleans cotton, in fact, did not reach and maintain the ten cent level until after 1850.[12]

Most Mississippi cotton growers did not feel the full weight of the depression until after the collapse of the cotton market in 1840. Subsequent to that date, however, many planters of the state, especially those who were debtors at the time of the crash, were forced into bankruptcy by the ever-narrowing gap between fixed costs of cotton production and steadily declining prices.[13] In the early 1840s law suits arising from debt became so common in Mississippi that, according to one observer, "some of the lawyers had their declarations in assumpsit printed by the quire, leaving blanks only for the name of the debtor, creditor and amounts." [14]

An indeterminable but large number of the less honorable and more energetic of the stricken planters managed in this period of distress to avoid foreclosure on their valuable movable property by stealing away to Texas with slaves, horses and mules. If accomplished successfully, illegal departures of this kind required luck, skill, and a certain amount of advance preparation. Lights were left burning in the "big house" and in the cabins of the "quarter" on the night of an exodus in order to deceive creditors or their agents who might have been on guard against the flight of a hard-pressed debtor. The Negro slaves were mounted upon the horses and mules, and in company with the owner they rode away under cover of darkness in an attempt to cross the county line before daybreak. If such fugitives were able to reach an adjoining county undetected, they were able to make the remainder of the journey to Texas in comparative safety, unless the defrauded creditor and his friends took the law into their own hands. This they sometimes did, and running gun battles occasionally were fought between fleeing slaveowners and creditors bent upon re-possessing slaves. Once inside the Republic of Texas, however, the absconding slave-owner was safely beyond the reach of American law, and at liberty to resume his occupation of raising cotton without fear of interruption by a sheriff. Midnight departures for the lusty young republic to the west were so common in the early 'forties, according to a popular joke of the time, that all Southwestern sheriffs found it expedient to obtain stamps marked "GTT" with which to label the great mass of legal documents involving former residents who had "Gone To Texas." [15]

Farmers and planters who remained in Mississippi and who managed somehow to survive the extended economic crisis were able to keep their heads above the financial waters only by making radical alterations in their methods of farming. As a class, cotton growers then and later tried to offset lowered prices by producing more cotton with the same labor force and to practice thrift. Under the spur of necessity, they reduced importations of foodstuffs and manufactured goods, which had been formerly obtained from England and the states of the North and Northwest; and they attempted to produce many of these necessities upon their own plantations. Inasmuch as their greatest cash expenditure before 1837 had gone to pay for horses and mules, corn, salt pork, and Negro clothing, many Mississippi cotton producers tried to save by raising their own foodstuffs and some of their livestock and by manufacturing their own shoes, bedding and clothing. In short, during the depression Mississippi's farms and plantations tended to become self-contained units. The only things that planters continued to import in quantity from the Upper Mississippi Valley were mules, blooded cattle, sheep, hogs, and items that could not be fabricated locally, such as hempen cloth, baling rope, sheet and bar iron for making plows, farm implements of various kinds, machinery for gins and presses, hats, shoes, and some of the cloth for Negro clothing.[16]

The conservatism that had made Mississippi cotton growers slow to adopt agricultural innovations before 1837 was an early casualty of the depression. The old hostility of planters to economic change of any sort had been based partly upon a conviction that their farming methods were already as profitable as any that could be devised. But when cotton began to sell for less than half what it had brought during the prosperous years of the early 'thirties, many farmers and planters began to question the wisdom of depending almost entirely upon the cultivation of a single crop. In a speech in July, 1841, Francis Leech, a Lowndes County planter, expressed the general change in attitude toward experimentation. To an audience of cotton growers assembled at Columbus to form a county agricultural society, he said:

> The object for which we have assembled is one of no ordinary interest. For the first time in the annals of our country have the farmers met to adopt measures that would promote not only their individual interests, but advance the permanent prosperity of the State. . . . The fact of our meeting on this occasion proves that a change in our mode of agriculture is necessary; for if those around me were satisfied the system they now pursue could not be improved, they would not assemble to hear the customs they had cherished either questioned or assailed. But the desire manifested by the farmers of Lowndes [County] to retrace past errors, speaks volumes. . . .[17]

Pinched by depression, thinking farmers began to seek methods of increasing their income and reducing their operating costs. This new willingness to make experiments and to test improved implements and techniques soon brought about a near agricultural revolution in the Old Southwest. Indeed, modern agriculture in Mississippi had its beginning in 1839.

From the very beginning of interest in agricultural reform in Mississippi, political newspapers provided a certain measure of leadership for the movement. After the depression had opened the minds of cotton growers to new economic ideas, political journals served as one of the most important media for communicating those ideas to the community at large. Newspapers became virtual clearing houses for agricultural information of all kinds. They published letters written by local farmers, planters, and mechanics, and made a practice of republishing articles of interest to cotton growers that had appeared originally in the pages of out-of-state papers and periodicals. Many of Mississippi's newspapermen were agricultural reformers themselves, and they used their journals for advocating the organization of state and county agricultural societies as well as for urging the adoption of many and varied agricultural improvement schemes.

During the depression several journalists interested in agricultural improvement made attempts to establish periodicals devoted to the special needs of Mississippi cotton growers, but none of them was able to build up a circulation large enough to make his venture financially successful. In fact, most of these publishing projects never got beyond the planning stage, for advance sales of subscriptions did not bring in enough funds to defray initial publication costs.[18] It proved so difficult to finance enterprises of this kind in the Old Southwest that only four agricultural periodicals actually appeared in print in Mississippi in the years 1839-50: the *Mississippi Farmer* (1839-40) and the *Southwestern Farmer* (1842-45), both published at Raymond, Mississippi, by Nathaniel Greene North; the *Planter* (1845-?) published at Holmesville by H. S. Bonney and B. Lewis; and the *Southern Planter* (1842) published at Washington, Mississippi, by Samuel E. Bailey. Of these papers, the *Southwestern Farmer* was the only one that received enough financial support from Mississippians to permit its survival for more than a year. Even at its height, the *Southwestern Farmer's* subscription list was discouragingly short—never more than four hundred names.[19] In the depression years, only a few farmers or planters in Mississippi apparently could afford to spend $2.50 a year for agricultural magazines, even when these journals were devoted to solving the very problems that beset them.

The Raymond *Southwestern Farmer* was far more important to Mississippi agriculture during the years 1842-45 than its brief lifespan and restricted subscription list would indicate. Subscribers frequently passed their copies on to friends after reading the paper themselves; and many agricultural societies made it available to their members along with other agricultural periodicals from various parts of the nation. Furthermore, political newspapers made a habit of republishing articles originating in the *Farmer*—a custom universally followed by journalists in the days before the founding of the news gathering agencies. Thus the *Farmer* was able to influence directly or indirectly a reading public several times greater than its four hundred paying subscribers. And it is probable that nearly all the farmers and planters of the state were exposed, at one time or another, to opinions about methods of agricultural improvement expressed by the Raymond paper's numerous correspondents.[20]

Both of the periodicals published at Raymond by Nathaniel Green North—the shortlived *Mississippi Farmer* and its successor, the more profitable *Southwestern Farmer*—were modeled after the leading agricultural journal of the early 1840s, the Albany (New York) *Cultivator*, founded in 1836 by Judge Jesse Buel.[21] Notwithstanding the poor financial support given them by the Mississippi agricultural community, both these pioneer Mississippi papers compared favorably with such other Southern periodicals as the *Soil of the South*, the *Cotton Planter*, and the *Southern Cultivator*. The editorial staff of the *Southwestern Farmer* was exceptionally qualified for the task. In addition to North, it included an able Mississippi journalist, John C. Jenkins, Jr.,[22] and, at a later date, Martin W. Philips,[23] a correspondent known to the entire American agricultural press of that day. As all three of these writers had been pioneer Southwestern reformers and were particularly active in the movement for establishing local and statewide agricultural societies, their contacts brought communications to their paper from most of Mississippi's leading farmers, planters, mechanics, and horticulturists. Consequently, the *Southwestern Farmer* was the almost official voice of Mississippi agricultural reformers during the period from 1842 to 1845.[24]

During the lifetime of his paper, North, as editor and proprietor, adhered firmly to the editorial policy enunciated in his prospectus.[25] In character the *Farmer* was a non-political family paper, devoted primarily "to our peculiar South-Western habits of farming." Under North's control, it advocated diversification and rotation of crops, the raising of fine livestock, the planting of fruits and grasses, the use of natural fertilizers like cotton seed and cowpeas, the use of improved farm tools, implements, and machinery, selective breeding of cotton and corn, and control of soil erosion by horizontal plowing and systems of guard and drainage ditches. North and his associates also encouraged the formation of agricultural societies, the building up of home industries, the founding of a system of popular education, and the temperance movement. Because the Raymond publisher went to great pains to secure the services of practical correspondents in order to make the *Farmer* a "medium of interchange of thought between agricultural writers from different parts of the state," his paper became a mine of information about agricultural practices and theories then current in Mississippi.[26]

The *Southern Planter,* published at Washington, Mississippi, by Samuel E. Bailey, as "the official organ of the Agricultural, Horticultural and Botanical Society of Jefferson College" was decidedly inferior in content, arrangement of material, and editorial quality to its contemporary, the *Southwestern Farmer.*[27] Bailey, who became mentally deranged a few years after settling in the Natchez area, was so incompetent that his venture into agricultural journalism was foredoomed to failure. He managed to publish eight issues only because he received a subsidy from the agricultural society of Jefferson College.[28] The *Southern Planter* at its peak had fewer than two hundred and fifty paying subscribers, and its influence upon the agricultural movement was negligible.[29] Nevertheless, the eight issues that appeared in 1842 are still of some interest because they contain detailed information about the activities of the Jefferson College society and a few valuable communications from farmers and planters of the Southern part of the state.

No copies of Mississippi's fourth agricultural newspaper, the *Planter,* published briefly at Holmesville in 1845, seem to have survived to the present time. The few known facts regarding it are gleaned from scattered references in other papers of the state. Because the *Planter* was seldom quoted by other newspapers, it seems reasonable to assume that it had little importance to Mississippi ante-bellum agriculture.[30]

During the depression years, the agricultural press of the North and East came to play an increasingly important role in Mississippi. Yet the leading Northern agricultural journals had not been completely unknown to Southwestern cotton growers in the pre-depression period. Jesse Buel's Albany (New York)

Cultivator, for example, had reached as many as fifty-four Mississippi subscribers in 1837.[31] But cotton planters at the time generally agreed that Northern farming practices were seldom applicable to the extensive cultivation of Southern staple crops.[32] Correspondents of Northern papers were writing then (as they did later) on such subjects as the cultivation of various types of grains, the building of pastures, breeding of purebred horses, cattle, sheep and hogs, fertilizing with animal manures, rotation of crops, deep plowing, drainage systems, fencing, and other matters pertaining to the intensive type of farming practiced in that section of the United States. These topics did not interest agriculturists of the Lower South, who were concerned only with raising as much cotton as possible within the limitations imposed by the size of their labor force. Their attitude changed during the depression, however, when cotton raising became comparatively unprofitable. Then Southwesterners became interested also in stockbreeding, horticulture, grain cultivation, and methods of conserving the soil—topics on which they were able to find much information of value in the pages of the Northern and Eastern agricultural press.

Information supplied by out-of-state agricultural periodicals became available to Mississippians during the depression years mainly through two agencies. One was the local agricultural society, which usually subscribed to the leading Southern and Northern farm journals. The other was the political press, which commonly reprinted agricultural articles of interest and merit. By the simple device of exchanging subscriptions, most political newspapers obtained a surprisingly large selection of the nation's agricultural periodicals. To cite two examples: the Natchez *Free Trader*, in 1842, was exchanging with the Albany (New York) *Cultivator*, the Rochester (New York) *New Genesee Farmer*, the *Connecticut Farmer*, the Nashville *Agriculturist*, the Charleston (South Carolina) *Southern Agriculturist*, the Augusta (Georgia) *Southern Cultivator* and the Cincinnati *Western Farmer and Gardener*.[33] The Jackson *Southron*, in 1843, was receiving the same journals and, in addition, was getting the Petersburg (Virginia) *Farmers' Register*, the Boston *Cultivator*, the *Maine Farmer*, and the *Central New York Farmer*.[34] The two Mississippi periodicals in this period—the *Southwestern Farmer* and the *Southern Planter*—were able, of course, to obtain files of practically all of the farm papers and magazines then appearing in the United States.[35]

From 1843 to 1860, by far the most widely read and most influential agricultural paper in Mississippi was the *Southern Cultivator*, published at Augusta, Georgia. In those years Mississippi editors continued to praise it and to quote from its columns; and the paper itself enjoyed a comparatively large circulation among farmers of the state.

The Augusta paper was edited successively by Jethro W. Jones, James Camak, and the nationally famous agricultural writer, Daniel Lee—all of them able men; and it was aimed particularly at the cotton growers of the Lower South. The editors sought to increase the profits of cotton growing through improved methods of cultivation and through substitution, wherever possible, of draft animals and improved horse-drawn implements for expensive manpower. At the same time, the *Southern Cultivator* advocated local and regional economic self-sufficiency as the best solution for the Lower South's agricultural problems. In this connection, the *Cultivator's* editors urged cotton farmers to produce their own foodstuffs, livestock feeds, and animals, in order to reduce their expenditures of cash. More important for the future of Southern agriculture, the *Cultivator* was a consistent critic of the Southern habit of destroying the soil by exploitive cultivation and then moving on to repeat the process in a new location. It pleaded for conservation of the soil by means of an organized system of crop rotation and the use of such natural fertilizers as cotton seed and cowpeas; and it aimed at checking the effects of erosion by means of contour plowing and systems of guard ditches and drains. The editors encouraged farmers to take advantage of recent advances in science and industry and urged the establishment of agricultural schools to train new generations of cotton farmers in the fundamentals of their occupation. On the other hand, Lee and his associates were wise enough to avoid the controversial topics of politics and slavery that had ruined other Southern agricultural periodicals. They did not attempt seriously to exchange Southern extensive cotton growing for the Northern system of diversified small farms. For these reasons, the *Southern Cultivator's* general editorial program was tailored perfectly to fit the prejudices and requirements of the average cotton grower in Louisiana, Mississippi, and Alabama, as well as in the paper's home state of Georgia.[36]

The collapse of the cotton market in 1840 had other effects upon Mississippians beyond disillusioning them with their great white staple. Contrary to their traditions of individualism, it taught them to join hands with their neighbors in solving common problems.[37] At the urging of many newspaper editors, groups of farmers in every part of the state began to band together in county agricultural societies. In so doing these farmers and planters were consciously following examples set in Virginia, South Carolina, and Georgia, where similar farm organizations had combatted depression by encouraging reforms.[38] Many of the meetings of these Mississippi societies were devoted to the discussion, not of works of leading Northern or European agriculturists, but of the writings and theories of George Washington, Thomas Jefferson, John Taylor of Caroline, Edmund Ruffin, and other Southern agricultural reformers. This partiality for Southern writings had little to do with Southern nationalism. Mississippians chose the works of Southeastern reformers simply because farm problems in the South Atlantic states were much like those in the Old Southwest—Northern and European writers were unfamiliar with the problems of slave labor and extensive clean culture of staple row crops.

The movement to organize local agricultural societies spread rapidly and almost spontaneously in Mississippi after its beginning in 1839. Oddly, the movement had two separate and independent places of origin: one, as would be expected, was in the old and highly cultured Natchez region. The other was in the comparatively recently settled Raymond-Clinton-Canton triangle of Hinds and Madison Counties, in the central part of the state.

In the late 'thirties, the idea of forming an agricultural society was no novelty to the people of Natchez. As early as 1819 one editor of that city had published a series of articles criticizing the local practice of depending entirely upon cotton, and calling for diversified farming and the formation of agricultural societies on the Virginia model.[39] Although cotton prices were then depressed, they were not low enough to compel Mississippi cotton growers to change from staple agriculture to the system of self-sufficiency advocated by Eastern agricultural societies. At no time before 1837 did farmers in the Natchez District exhibit any interest in organizations that urged reform. Yet, during those years, the Mississippi press kept planters in the Southwestern part of the state informed about contemporary developments in the North and in the tobacco and cotton raising states of the Atlantic seaboard. Consequently, most cotton growers of the Natchez region were adequately prepared to follow the example of older sections of the South when the need for agricultural reform in Mississippi eventually became pressing.

From the time of its founding in 1802, Jefferson College at Washington had been the intellectual center of the old Natchez District.[40] It was natural, therefore, for the state's first agricultural society to be closely associated with the science department of that institution. Furthermore, most of the active sponsors of the new society were warm friends of the college. The first organizational meeting of the Agricultural, Horticultural, and Botanical Society of Jefferson College assembled on May 6, 1839. The twenty-five charter members were mostly trustees and professors of the institution, plus several wealthy slaveholding cotton planters from the neighborhood of Natchez. They designated the college auditorium as the society's permanent meeting place, and subsequently held many public exhibitions on the college grounds.[41]

The constitution of the Jefferson College society proclaimed that its primary purpose was to improve economic conditions through agricultural education.[42] Believing, like all other Southern reformers of the depression period, that over-concentration upon cotton was at the root of their economic difficulties, the members worked diligently for several years to foster regional self-sufficiency in the Old Southwest. Accordingly, they advocated home manufacturing and the creation of local industry, the raising of purebred horses, cattle, sheep and hogs, and the cultivation of fruits, grasses, and grains. During the early depression years, they carried on their educational program by every means at their disposal.[43] They welcomed the general public to meetings of the society, where members lectured on the history of agriculture and on improving farming methods, and where everyone could participate in discussions on common problems. In 1842 the society undertook to bring its program before wider audiences by sponsoring publication of the ill-fated *Southern Planter*.[44] Fairs and exhibitions, however, were the society's most effective means of influencing the thinking of Southwesterners.

Directly after it was organized, the Jefferson College society began to hold modest exhibitions of fruits, flowers, and vegetables raised by members and their families. These simple horticultural shows were soon expanded into small local fairs featuring the best specimens of imported and native livestock of all kinds to be found in the neighborhood of Natchez. These fairs also demonstrated locally-made tools and implements and a few improved devices of Northern manufacture; but it was the livestock shows that attracted by far the greatest attention between 1839 and 1843. In those years something like a mania for thoroughbred cows, sheep and hogs swept through the cotton growing sections of Mississippi, and this new interest in fine animals brought farmers and planters flocking to the Jefferson College agricultural fairs.[45]

The Washington society was able to provide livestock shows of unusual interest in 1841 and 1842 largely because a number of wealthy planters had recently imported thoroughbred cattle and hogs on a rather lavish scale from Europe and the Northern United States. Thomas Hall had taken the lead in introducing the famous Durham breed of beef cattle into the neighborhood; and a magnificent three year old bull named "Beltzhoover," weighing 2300 pounds, which Hall had purchased in Baltimore, was the center of attention at fairs held in Adams and Jefferson counties in these years. Dr. James Metcalf, another planter of Natchez, had imported a shipment of Ayrshire cattle directly from England with the object of improving the milking qualities of the native "neat" cattle of South Mississippi. His stock of purebred and halfblood animals were also important attractions at the Washington livestock exhibits. Furthermore, in 1841 and 1842 several livestock dealers of the Upper Mississippi Valley brought fine cattle, usually of the Durham variety, down the river in order to display them for sale at the fairs of the Natchez District. In one instance of this kind, Thomas Affleck, an editor of the Cincinnati *Western Farmer and Gardener,* purchased some English Durhams in Ohio and brought a number of bulls and heifers to Mississippi for resale. When his best bull, "Cincinnatus," proved to be too expensive for the purses of individual Natchez planters, several of them banded together to purchase the animal on shares. Unfortunately this fine beast, like many other imported cattle of the Durham breed, was unable to survive very long in the hot climate and rough pastures of South Mississippi.[46]

Thoroughbred hogs attracted almost as much attention as cattle at the Natchez fairs during the early depression years. Berkshire pigs, which had been introduced into the state in the late 1830s, rapidly won popularity because of the great size they attained at comparatively early ages. Numerous specimens of this breed, in both the black and the white varieties, were always featured at the Southwestern fairs, and they won such universal esteem that they became ante-bellum Mississippi's most popular breed of hog. Berkshires, however, were by no means the only variety displayed. There also were many specimens of Skinner's No-Bone breed and a few Neopolitan pigs belonging to Thomas Hall, as well as an assortment of Irish Graziers, Woburns, Bedfords, and numerous crosses between these breeds and the native Land Pikes of the Natchez District.[47]

Despite the suitability of the Natchez region for sheep-raising, few sheep were exhibited. The occasional specimens that were displayed were usually undistinguished crosses between the Spanish Merino or the English Bakewell and the common breed of Southern Mississippi. Inasmuch as most planters were interested in raising sheep primarily for meat rather than for wool, they were generally satisfied with this mongrel animal, which produced an excellent grade of mutton but practically no fleece.[48]

Although fine cattle and hogs attracted much attention at

the Washington fairs, keen-eyed observers could have seen that most planters were more interested in fine race horses, saddle horses, and carriage horses than they were in superior breeds of cattle, sheep, and hogs. The interest in fine livestock, which had bidden fair in 1840-42 to grow strong among the wealthier planters of the Natchez region, soon began to wane, and their concern with general agricultural reform ebbed away. Consequently, the livestock exhibits of the Jefferson College agricultural society were discontinued in 1843, and horse shows and horse races took their place in the Natchez District.[49]

This decline of interest in the Washington livestock shows was merely symptomatic of flagging interest in the entire field of agricultural improvement. Even though cotton prices were sinking lower each year, by 1843 most planters had managed to reduce their operating costs to a level corresponding with the lower prices they received. They had increased their cotton production by more efficient farming methods. By raising their own food and by spinning and weaving their own cloth, they supplied most of their own necessities. Having successfully adjusted their agricultural methods to prevailing depression conditions, cotton growers tended to lose interest in organizations dedicated to further improvement. Indeed, the conclusion is inescapable that the leaders of the Jefferson College society had exhausted their intellectual resources in matters of reform. The addresses delivered before the society in 1843 were notably lacking in new ideas, and exhibitions of that year were uninspired repetitions of earlier livestock shows. In 1843 Thomas Affleck, a Scottish agricultural expert who had settled at Natchez in 1842, conducted plowing competitions at the Washington fair to demonstrate the value of improved Northern plows, but these tests attracted little or no attention from large slave-owners who were accustomed to leaving such details to their hired overseers.[50] During 1843 attendance at the meetings of the society grew steadily smaller in spite of energetic efforts on the part of Affleck and of B. L. C. Wailes, the perennial president of the organization. By late 1845 the society was practically disbanded.

The Agricultural, Horticultural and Botanical Society of

Jefferson College failed to become a permanent force for improvement mainly because of the limitations of its members. With few exceptions, they were wealthy cotton growers who had been trained in law or in the liberal arts, and their primary interests were directed toward their social life and the political affairs of the state rather than toward the practical operation of their plantations. Almost all of the members of the society were large slaveowners who left the management of their plantations mainly to their white employees, and it is probable that they knew comparatively little of the technical side of cotton growing. With the sole exception of Thomas Affleck, none of them was willing or able to conduct the systematic experiments necessary for basic agricultural advancement. Consequently, the Jefferson College society was given more to discussion and theory than to practical experimentation, and its fund of knowledge was exhausted after a few meetings.[51]

The movement for agricultural improvement that originated at Raymond in June, 1839, followed a somewhat different course. Yet it too eventually came to grief from the same causes that brought about the demise of the society at Washington. Where the latter organization was concerned mainly with the state's southwestern corner, the Raymond group aimed from the start at revolutionizing the whole of Mississippi agriculture by building up an integrated statewide organization of county societies.[52]

The Raymond reform movement was initiated by Martin W. Philips of Edwards, a small planter who had been an avid reader of Northern and Eastern agricultural journals ever since his arrival in Hinds County in 1831. Writing in the Raymond *Times* under the pseudonym of "Pro Bono Publico," Philips, on June 6, 1839, issued an appeal to the planters of Hinds County to convene at the county seat for the purpose of organizing an agricultural society.[53] North, the *Times'* editor, seconded Philips's plea for such an organization. In an editorial he pointed out that other sections of the country had improved their economic condition through livestock raising, diversification of crops, and other reforms which had been initiated under the auspices of agricultural societies.[54]

The appeals made by Philips and North struck responsive chords in the ears of planters of Hinds and Madison Counties. In June and July they founded societies at Clinton, Raymond and Canton. As soon as duplication of effort became known, the Raymond and Clinton organizations began to discuss consolidation. Eventually on December 23, 1839, they merged into one body called the Agricultural Society of Hinds County. In the meantime, both organizations cooperated fully with one another in efforts to form a state organization.[55]

At the recommendation of John C. Jenkins, Jr., the members of the Raymond Society went on record on June 24, 1839, as favoring the establishment of a state agricultural society. They urged the various county organizations to send delegates to a meeting in Jackson. In order to gain support for the project, the Raymond reformers instructed a committee to write a manifesto calling upon the people of the state to form county societies as auxiliaries of the proposed state agricultural organization.[56]

On August 2, 1839, this appeal was published in the Raymond *Times* and was subsequently reprinted in whole or in part by many of the newspapers of the state. Lengthy though it was, the article aroused great interest among planters all over Mississippi. The committee attributed the state's pressing economic problems to the exclusive growing of cotton, and urged the people to make themselves economically independent. In this proclamation, the farmers were advised to produce their own food, raise their own livestock, manufacture their own clothing, and preserve the fertility of their soil. The manifesto, declaring that the economic ills of Mississippi were statewide, not merely local, called upon the planters and farmers of the state to band together in county organizations in order to attack their difficulties in unison; and it urged local societies to coordinate their labors through a state society.[57]

In response to the Raymond manifesto, a number of county societies were organized in various parts of the state during the summer and fall of 1839, and most of them made provisions in their constitutions for taking part in a state society whenever it should be established.[58] By common consent leadership in setting up the state organization was left to Raymond. When the Raymond and Clinton groups merged to become the Hinds County Agricultural Society in December, 1839, the members issued an invitation to all local societies to send delegates to a meeting in Jackson to be held during the approaching session of the state legislature.[59] They themselves chose a delegation of fifteen members, including N. G. North, John C. Jenkins, Jr., and the noted stockbreeder and Democratic politician, Colin S. Tarpley, to represent Hinds County at this meeting.[60]

On the evening of January 22, 1840, the meeting called by the Hinds County society convened in the representatives' hall of the State Capitol.[61] Attendance was large because many persons from all parts of the state had been drawn to the city by the regular session of the legislature. A constitution for a state agricultural society was adopted that night, officers were elected, and committees appointed. Adam Bingaman of the Jefferson College society was chosen to be the featured speaker at the next annual meeting. Although those present displayed much enthusiasm for the cause of agricultural improvement, their ardor was ephemeral; it was probably a product of the inspired oratory of the evening, not the result of sincere interest in the subject. At any rate, only a few of the participants were willing to exert themselves after the initial effort. During the next two years the organization undertook no activity. At the time appointed for the second annual meeting in 1841, not a single officer or member appeared.[62]

The local societies likewise suffered from the prevailing attitude of indifference, and nearly all of them ceased holding regular meetings during 1840 and 1841.[63] Even the Hinds County organization, which had been so active in the reform movement during 1839, was affected by the general lassitude. A few of the original members refused to become discouraged at the lack of concrete results, but it was mainly as individuals that they continued to work doggedly for their cause.[64]

In a desperate effort to generate support for their program of livestock raising and diversified farming, a number of the Hinds County reformers, including N. G. North, Martin W. Philips, Colin S. Tarpley, Gaston Kearney, John B. Peyton, Henry Andrews, Thomas Hudnall, Eli T. Montgomery, and R.

M. Williamson, decided to hold a livestock show at Raymond. This event, the first local fair held in Central Mississippi, eventually took place on December 19, 1841. The principal exhibits were a large number of thoroughbred Berkshire hogs, some Durham cattle belonging to Tarpley, Philips, and Peyton, a pair of very fine imported Southdown sheep from Tarpley's flock, and a small drove of Durham cattle brought to Raymond for sale by a Quaker livestock dealer from the Upper Mississippi Valley. Many farmers and planters of the neighborhood attended the small fair with their families, and most of them appeared favorably impressed by the superior merits of the blooded stock they saw. Yet when they inquired into the possibility of purchasing animals similar to those in the show, they discovered in most cases that the prices of thoroughbred cows and hogs were far beyond their reach. Nevertheless, the Quaker dealer did dispose of a few Durhams to spectators at the fair. Furthermore, this fair was directly responsible for stimulating an interest in good stock among Hinds County farmers, which eventually caused some of them to purchase thoroughbred or crossbred Berkshire hogs and Durham cattle. Within a few years after the Raymond fair, Hinds County was able to boast as fine a collection of blooded stock as any other part of the state, including the wealthy Natchez region.[65]

After lying dormant in 1840 and 1841, the movement for reform took on new life in 1842, and enjoyed a brief but energetic career that year and the next before relapsing into a coma during the latter half of the decade. Several factors were responsible for this revival. One was the extensive publicity given to a national farm organization at the time of its founding at Washington, D. C., in December, 1841. A second was the organization of state agricultural societies about this time in South Carolina, Georgia, Alabama and Louisiana. Still another stimulus came from numerous state and local fairs then being held all over the cotton producing portions of the South. But the most influential factor of all was the establishment of the Raymond *Southwestern Farmer,* a competent paper which gave Mississippi enthusiasts a convenient forum for communicating their ideas on agricultural reform.

This renewed interest in improvement led to the formation of numerous new county societies in the years 1842-46, and some of the older ones came back to life.[66] Moreover, the state society, which had remained inert after its establishment in January, 1840, became active and began for the first time to hold annual meetings at Jackson.[67] As a direct result of renewed activity among reformers, a large number of state and county fairs were held in 1842, 1843, and 1844 under the sponsorship of various agricultural organizations. These events acquainted cotton producers with the latest developments in agriculture and horticulture.[68] They also gave farmers opportunities to examine many of the recent improvements in farm implements and machinery.

All these fairs were of the same type, whether they were conducted by local organizations or by the state association. Their primary object was to emphasize that it was essential for farmers and planters to produce as much of their necessities on their own farms and plantations as possible, thus reducing their outlays of cash. With this purpose in mind, agricultural reformers delivered speeches and lectures to the crowds assembled at the fairs, and the sponsoring agencies awarded prizes of various kinds for the best specimens of blooded horses, cattle, sheep, and swine, and for the best samples of cotton, corn, small grains, fruits and vegetables produced in the state. As the promoters of fairs were interested in regional as well as in individual self-sufficiency, they frequently offered prizes and certificates of excellence for the best specimens of such locally manufactured goods as plows, gins and cotton presses, Negro hats, shoes, clothing, and bedding. At these gatherings, the livestock shows ordinarily attracted the greatest crowds and aroused the most discussion. Fair-goers generally gave least attention to farm implements and machinery. Their indifference may have been due to scanty exhibits of machinery. On the other hand, the meager exhibits may have been a result of the public's indifference.[69]

Because agricultural fairs were recognized as having social as well as educational functions, program committees always included events of special interest to farm women. Contests were held for the best specimens of home-cooked preserves, pies, cakes, smoked hams, bacon, wines, and other edible products of the kitchen, garden and dairy. Competitions also took place to determine the best examples of needlework, floral arrangements, and many other types of feminine workmanship.[70] Unfortunately, the fairs of the 'forties gave little attention to the need for providing other kinds of public entertainment. Consequently, the spectators tended to become bored after seeing the same things at successive exhibitions, and the size of the crowds always began to diminish after several fairs had been held in the same neighborhood.[71] In contrast to the fairs of the 1840s, those held during the latter 1850s attracted increasingly large audiences of holiday-minded farm folk. This popularity resulted from the addition of tournaments, horse races and other forms of public entertainment to the usual exhibitions of livestock, implements and farm produce.

Notwithstanding their failure to retain the interest of the farm community for more than a year or two, the state and county fairs of the early 1840s were not without beneficial results. Farmers received opportunities to examine specimens of the thoroughbred livestock which they were reading about in agricultural and political newspapers. They also could purchase breeding stock at fairs if they were minded to improve their cattle, sheep, and hogs. Indeed, many cotton farmers and planters did take advantage of the chance to buy fine stock. The thoroughbred bulls, bucks, and boars they purchased markedly improved the quality of Mississippi's flocks and herds during the depression decade. These fairs also encouraged the modernization of farm equipment to some extent. Spectators at these events occasionally purchased improved implements and machinery, and the consequent adoption of better plows, gins, presses and mills, along with newly developed cultivators, cotton scrapers, double shovels and other specialized horse-drawn farming implements, resulted ultimately in a remarkable increase in the productivity of Mississippi farm labor. Finally, cotton growers gained varying but undoubted benefits from exchanging ideas on agricultural matters with friends, neighbors and new acquaintances at these gatherings.

Because organized agricultural reform in Mississippi had been a product of financial hardship, its decline and extinction coincided significantly with improving economic conditions in the Old Southwest. When cotton prices began to rise slowly during the latter half of the decade, the farmers—who had learned to manage successfully even when cotton brought only five cents a pound—began to regain a semblance of their former prosperity. They lost interest in making further modifications in their system once they began to make money again. Self-sufficiency had no charms in itself!

Between 1845 and 1850 only a few county agricultural societies were founded. In 1845 the state organization and its affiliated local associations lapsed into inactivity in spite of vigorous protests from Thomas Affleck, Martin W. Philips, John C. Jenkins, Jr., Nathaniel Greene North, and other leading agricultural reformers. These leaders continued to urge that cooperation between all cotton farmers and planters was essential for avoiding eventual impoverishment as a result of soil erosion and diminishing soil fertility.[72]

The collapse of the state and county associations in 1845 did not indicate that Mississippians were reverting to their old system of cultivating almost nothing but cotton, or that they were willfully closing their eyes to further scientific progress. Indeed, quite the reverse was true. Agricultural publications from all parts of the Union circulated in ever increasing numbers through the Mississippi farm community during the remainder of the ante-bellum period, and the local papers found that their subscribers looked with favor on agricultural articles. Hence it was not difficult for Mississippians as individuals to keep informed about the latest improvements made by agriculturists in Europe and the United States.[73]

During the years between 1845 and 1850, large numbers of the state's more progressive farmers continued to improve the quality of their livestock through selective breeding, introduction of new blood lines, and building better pastures. In those same years a majority of cotton growers continued to minimize their costs of operation by raising most of the food needed for their farm workers and livestock. Many of the larger planters also increased the productivity of their laborers by providing them with improved plows, cultivators, and other kinds of farming equipment, along with additional teams of horses and mules. For the first time, orchards of apples, peaches and pears were planted on numerous plantations and farms of Central and Northern Mississippi. Even more important, improved varieties of cotton, developed by applying the principles of selective breeding to the cultivation of the old Mexican and Petit Gulf varieties, greatly enhanced the quality, if not the quantity, of the staple produced in the state. In short, Mississippians had not lost their concern for superior methods of farming in the late 'forties. Even though leading farmers had dropped out of agricultural societies, they in many cases had merely rejected the cooperative approach to economic problems, and were dealing with them as individuals instead.[74]

CHAPTER   V

Livestock and Livestock Breeding

An abundance of contemporary evidence leaves no doubt that the 1837-49 depression made an indelible imprint upon the agricultural history of Mississippi. Indeed, a decade of subnormal cotton prices and the agricultural reform movement which they had engendered made lasting and fundamental alterations in the economic life of the state. In the words of one cotton planter, written in 1841: "So great a change has occurred in the manners and habits of the people, as well as in their domestic policy, that I can hardly assure myself of the fact that I am in the midst of the same community and a dweller on the same soil that received me long years since."[1]

King Cotton was not dethroned by his Mississippi subjects during these years, but he was reduced from the status of an absolute despot to that of a constitutional monarch. Cotton continued to be the principal source of farm income, but other crops began to receive proportionately larger shares of the available supply of labor. On many plantations in the early 1840s, livestock breeding, development of pasture lands, cultivation of corn, small grains, and other foodstuffs, planting and management of orchards, and conservation of the soil became almost as important as the production of cotton itself. Even after the return of good cotton prices in the late 1840s, these depression-born pursuits were not discontinued, although, to be sure, their share of the average cotton grower's attention was somewhat diminished.[2]

While Mississippians had raised corn, beef and pork in con-

93

siderable quantities before the depression, they had always consumed more of these products than they had grown. The deficit they had easily made up by purchases from the Upper Mississippi Valley. In fact, during the early 1830s when cotton prices ranged between fifteen and twenty cents a pound, cotton growers had generally considered it an economical practice to purchase corn at thirty-seven to fifty cents a bushel and salt meat at from three to four cents a pound. Before 1837, therefore, hundreds of thousands of dollars had left Mississippi annually to pay for produce bought in the Upper Mississippi Valley.[3]

When the depression arrived, cotton declined in value more rapidly than Northwestern products; and the widening differential compelled the cotton growers of Mississippi to reduce their cash expenditures for imported foodstuffs and manufactures. What tumbling cotton prices meant to the agricultural community of the state was illustrated by some computations made by a committee of the Raymond agricultural society in August, 1839. Assuming that the average net profit from a bale of cotton at current prices was forty dollars, the committee estimated that a barrel of salt meat purchased at the river bank would cost a planter living twenty miles in the interior the equivalent of two-thirds of a bale of cotton, if the cost of handling were added to the average purchase price of twenty dollars a barrel. This cost was almost pure loss, for Land Pike hogs were self-supporting, and to turn them into cured pork cost their owners little except for salt and the labor involved in processing the meat. The only sound course of action for Mississippians, the committee concluded, was to produce their own supplies of corn, pork and other foodstuffs. As most planters of the state were of the same mind, livestock breeding and meat packing developed into important domestic industries during the first phase of the depression.[4]

When Mississippians began to undertake pork production with real earnestness, the Land Pike was the hog most frequently found in Southwestern pens and woodlots. But changes in breed were not long in coming. Shortly after deciding to supply their Negroes with home-grown meat, a few of the better informed planters became convinced that "the old, pirating, fence-breaking, corn-destroying, long-snouted, big boned and leather-bellied" Land Pike hog was not ideal for their purposes.[5] From experience they learned that the free-foraging "Alligator," though hardy enough, was small, slow in maturing, difficult to fatten properly, and yielded an inferior quality of pork.[6]

Having once become dissatisfied with the Land Pike's pork-making qualities, several like-minded planters in different parts of the state began to experiment independently with imported thoroughbred stock. Their aim was to introduce or develop a breed of swine that would grow larger, fatten faster, and make a better grade of meat than the common "Alligator," and at the same time retain some of its hardiness.

William J. Minor, of Natchez, appears to have been the first to introduce thoroughbred swine into Southwestern Mississippi. Minor is better remembered as a sportsman and breeder of race horses than as a scientific agriculturist, but among other activities he did find time before 1839 to import Irish Graziers and White Berkshires from England. On his plantation, the Graziers and Berkshires were interbred, and boars resulting from this mixed parentage were mated with common Land Pike sows. The eventual result of this multiple crossbreeding was a large white porker that became famous locally as the "Minor" hog. Other planters of the area obtained specimens of the Minor stock and used them to improve the quality of their own herds.[7]

After a few years' popularity in the Natchez area, the Minor hog was superseded by thoroughbred animals obtained from Europe and the Northern United States. In the years 1840-42 Thomas Hall was the leading importer of blooded swine in the Natchez District, although many other planters were also purchasing improved pigs from various sources. He imported Neapolitan stock from Italy, Irish Graziers, Woburns, and Black Berkshires from England, and Skinner's No Bone hogs from Baltimore.[8] Hall followed Minor's practice of crossing pure-blooded hogs in various combinations, and some of his experiments produced happy results. One of his boars, Tam O'Shanter, the offspring of a Neapolitan sire and a Black Berkshire dam, became widely known during the early depression years. At the age of nine months Tam O'Shanter was the undisputed star of

the spring fair of the Jefferson College agricultural society, and
he retained his pre-eminence at subsequent livestock shows.[9]
When finally butchered, he was found to weigh nearly eight
hundred pounds.[10]

Thomas Affleck, the Scottish-born horticulturist and reporter
for the Cincinnati *Western Farmer and Gardener,* was another
important contributor to livestock improvement in Central and
Southern Mississippi during the early 1840s.[11] While reporting
Kentucky fairs and livestock shows for his paper in 1841, Affleck
as a speculating venture purchased ten head of choice Durham
cattle and some forty head of pedigreed Berkshire, Irish Grazier
and Woburn hogs. At the suggestion of Martin W. Philips,
Affleck then took these animals down the Mississippi River to
put them on sale in the markets of Vicksburg, Natchez and
Baton Rouge. Some twenty or thirty planters were on hand to
greet him on his arrival at Vicksburg in December, and they
forthwith bought more than half of his stock of hogs.[12] Leaving
some of his remaining animals in Vicksburg under the care of
Philips, Affleck proceeded on to Natchez. When he arrived there
a fair and livestock show sponsored by the Jefferson College
agricultural society was in progress at the neighboring town
of Washington. Affleck placed his animals on exhibition at the
fair, and several of them were purchased by local planters.[13]

The agricultural journalist was so pleased with the friendly
reception accorded him at Natchez and so favorably impressed
with the soil and climate of the neighborhood that he decided
to resign his position with the *Western Farmer and Gardener*
and make his permanent home in Southwestern Mississippi. A
local lady doubtless had something to do with Affleck's decision;
for he soon married and settled near Washington upon a small
farm owned by his bride. At "Ingleside," the home of the
Afflecks until the mid-1850s, his livestock that had remained un-
sold after the fair became the nucleus of small but noteworthy
herds of thoroughbred cattle and hogs.

The experiments in breeding conducted by Affleck, Hall,
Minor, and other members of the Jefferson College agricultural
society wrought decided improvements in the swine of Southwestern Mississippi during the latter years of the depression.

A Wilkinson County planter in 1842 observed: "The long snouts, slab sides, and porcupine backs of our land pike and alligator breeds are fast being superseded by the neat, full and elegant contour and filling up of the Berkshires, the Woburns, and the Graziers. . . ." [14]

Although most of the breeds of thoroughbred hogs in the United States at the time were represented in the herds of the old Natchez District, the Black Berkshire by the late 'forties was fast becoming the favorite not only there, but also in the central and northern sections of the state. [15] The trend in favor of the Black Berkshire was apparent as early as the spring of 1843, when hogs of that breed received most of the honors awarded for swine at the April livestock show of the Jefferson College society. [16]

In Central and Northern Mississippi, no one in the 1840s did more to encourage the raising of fine hogs, sheep and cattle than Martin W. Philips. Dr. Philips had migrated to Hinds County from his boyhood home in South Carolina in 1831. Although trained in medicine, he soon took up cotton planting, and moved in 1836 to a small plantation that he called "Log Hall," located about nine miles south of the town of Edwards on the bank of the Big Black River. Like many cotton growers Philips suffered losses from the panic of 1837 and the ensuing depression in the world textile trade. Financial reverses resulting from over-concentration upon cotton convinced Philips that the traditional agricultural methods of the cotton states were fundamentally unsound. For a decade or more he had been a careful reader of the leading agricultural journals; and it was from them that he got his ideas about causes and cures for Southern farm ills. The trouble, as he saw it, was simple: cotton farming in the old manner had involved spending too much money for imported commodities, especially in view of the modest income received from cotton. The solution, in Philips's opinion, was equally plain: improved techniques would increase cotton yields and would correspondingly raise planters' incomes. At the same time, costs of operation should be minimized by raising on the farm most of the food and livestock supplies needed. [17]

From the year 1839 onward, Philips worked diligently to promote this ideal of economic independence for farm, state, and region.[18] Pre-eminently an investigator and innovator, he tested new plows and farm equipment and carried out countless experiments in livestock breeding, crop diversification, and improving varieties of cotton; and the press cooperated in giving publicity to the results.

Philips was especially successful in his efforts to improve the quality of livestock in his own part of the state during the depression. He encouraged a number of Hinds County planters to go into the business of breeding and selling thoroughbred cattle; and his own experiments with sheep helped to dispel the common notion that wool could not be produced profitably in the warm climate of Mississippi. He achieved his greatest success, however, with thoroughbred hogs. Countless experiments with many different breeds and with various systems of caring for swine made him an acknowledged authority on the subject from Texas to the Atlantic seaboard.

In the early 'forties Dr. Philips conducted a large number of comparative tests with Woburns, Beltuses, Skinner's No Bones, Byfields, Chinas, Calcuttas, Irish Graziers, and Berkshires of both black and white varieties, with a view to determining which breed would gain the most weight and fatten most quickly when fed a measured quantity of corn.[19] These tests convinced him that Black Berkshires were ideal meat-producing hogs, for they could be counted on to reach 150 pounds at five months, 190 pounds at nine and a half months, and 290 pounds at eleven and a half months.[20]

Thereafter he made himself the most outspoken champion of Black Berkshires in the cotton belt. Many of his letters enumerating their merits and giving the results of his experiments on their feeding and care were published in the *Southwestern Farmer*, the *Southern Cultivator*, and other Southern newspapers and agricultural periodicals.

Northern and Eastern agricultural journalists like John S. Skinner and R. L. Allen purchased outstanding specimens of the Black Berkshire breed for Philips at fairs they covered for their papers, and shipped the animals to him at Log Hall. Philips then made the offspring of his pedigreed Berkshires available to the farmers and planters of the Vicksburg-Jackson region who wished to improve their herds of swine.[21]

Colin S. Tarpley and John B. Peyton, both of Hinds County, were almost as influential as Philips in improving the plantation hogs of Central Mississippi. Tarpley's plantation, "Hard Times," lay between Raymond and Jackson. There he specialized in breeding cattle and horses; but he was also deeply interested in both sheep and hogs. As early as 1840, his herds of swine contained many specimens of pedigreed Black and White Berkshires, and his flock of sheep was made up of thoroughbred Southdowns and Cotswalds (or Bakewells, as they were often called in Mississippi). Tarpley's livestock business was quite profitable; and in the early 'forties he shipped pigs to all parts of Mississippi, and even to buyers in Kentucky and Ohio.[22]

John Peyton, by contrast, was primarily interested in raising fine hogs. To his plantation near Raymond he imported the first of his pedigreed Black Berkshires in 1841 directly from the town of Berks in England. Two of his original hogs were noble specimens of their kind. Of these the boar, Prince Albert, weighed four hundred pounds at the age of two and a half years, and Albert's consort, Victoria, was only slightly smaller. In 1842 Prince Albert was the recipient of a unique honor. John C. Jenkins, Jr., the associate editor of the Raymond *Southwestern Farmer,* made an engraving of his Porcine Highness and published the picture in his paper. Thus Peyton's Prince Albert became the first Mississippi hog to have his likeness preserved for posterity.[23]

Dr. George Smith, of Warren County, was a good example of a Mississippi cotton grower who acquired a slight interest in breeding improved swine during the early years of depression.[24] Smith purchased a few Irish Graziers and Byfields from several dealers in 1842. After experimenting rather haphazardly with these thoroughbreds, Smith succeeded in breeding a variety of hog to his liking by crossing Black Berkshires with the hardy Land Pikes. Thereafter his mongrel Berkshire-Alligator stock formed the mainstay of his herds. Smith, like many of his contemporaries, did nothing, however, to improve his hogs beyond adding pureblood strains to their bloodlines. When Solon

Robinson visited Smith's Warren County plantation in 1848, he was unfavorably impressed by the doctor's method of raising hogs. The pigs were allowed to run free in the woods; and, until they were almost old enough to kill, they had no feed other than what they could find for themselves. A few weeks before butchering time, Smith's hogs were driven from the woods into wooden pens by gangs of Negroes and dogs. Once in the pens they were fed in troughs on boiled shelled corn and kitchen wastes. On observing that corn was selling for forty to fifty cents a bushel and salt pork for only three to five cents a pound at Vicksburg, Robinson wrote: "Which would be the best economy: to shoot the hogs and sell the corn and buy pork: or feed it, with the hope of making it of such hogs?" [25]

As North Mississippi was still in the process of settlement during the 'forties, it was affected later and to a lesser degree than Central and South Mississippi by the spirit that Colin Tarpley once characterized as "a perfect mania upon the subject of improved livestock." For example, in Carroll County, in the northcentral part of the state, no pureblooded sheep, cattle or hogs were to be found before 1841. In that year, however, John W. Kendall migrated from Kentucky and brought with him a large stock of Durham cattle, Berkshire, Leicester and Irish Grazier hogs, and sheep of the Southdown, Cotswald, and Saxony breeds.[26] Kendall was almost as enthusiastic about blooded livestock as Philips and Tarpley, and he soon infected a few of his neighbors with "the spirit of improvement upon the native stock." Two Carroll County planters, John A. Binford and Richard Eskridge, obtained Berkshire stock from Kendall shortly after his arrival in Mississippi and imported other animals from England, and both began to raise thoroughbred hogs on their plantations. From Carroll, the interest in fine animals created by Kendall's activities spread into adjacent regions, and small numbers of pedigreed livestock began to have visible effects upon the quality of herds in Holmes and Yalobusha Counties.[27]

The first group of Berkshires and Durhams to reach northernmost Mississippi were distributed among farmers in Lafayette, DeSoto and Marshall Counties in November, 1841, and the superiority of these animals was immediately apparent to all who saw them. A resident of Holly Springs declared hopefully: "Such incoming of improved stock into this portion of the State will do much to arouse public-spirited individuals to an enterprise in which this whole population are more immediately interested than any portion of the southwestern country." Yet such hopes were largely vain. Northern Mississippi did not become noted for fine livestock either during the ante-bellum period or for nearly a century thereafter. To begin the breeding of pedigreed cattle and hogs cost money: and the northern red clay hills seldom produced enough wealth to finance thoroughbred livestock farms.[28]

The same economic forces that generated interest in pureblooded hogs among Mississippians also gave rise to a desire to improve the common Creole cattle of the country. Wealthy Natchez planters who had supplied the leadership in hog breeding also furnished inspiration for this endeavor. Captain Isaac Ross of "Red Lick," according to a Natchez editor, was the earliest in the state to give attention to the improvement of cattle. The date of Ross's first importation of thoroughbred breeding stock is not known; but it doubtless was in the 1820s, for the Ross herd was famous locally at the time of his death in 1836.[29]

The next planter of the Natchez District who undertook the breeding of improved cattle on an important scale was Samuel Chamberlain of the Pine Ridge community. In 1835 he began by purchasing some heifers from Ross and a few other cows of the Patton breed from livestock dealers in Kentucky. Then he added to his herd a thoroughbred Durham bull from the stock of James Cage, a widely known Louisiana stockbreeder of the period. The mixture of these three strains ultimately produced large, well proportioned, healthy animals that thrived in the hot climate of South Mississippi. Chamberlain's "highly improved and beautiful cattle" were greatly admired by spectators at the Jefferson College fair in 1842, where they shared attention with thoroughbred stock belonging to Thomas Hall. Chamberlain's obvious success in breeding fine livestock undoubtedly was a major inspiration to his neighbors when the 1839 depression turned their attention to the need for economic self-sufficiency.[30]

It was Thomas Hall, however, who did more than any other man to develop an interest in fine livestock of all kinds among planters of the Natchez trade area. During the early depression years Hall spared no expense in importing prize winning Durham cattle from England and from the northern and western cattle raising districts of the United States. When Thomas Affleck visited the Jefferson College agricultural fair in December, 1840, he reported to the *Western Farmer and Gardener* that Hall's display of hogs and cattle was by far the most impressive that he saw. Hall's Durham bull, "George Beltzhoover" (purchased from the Baltimore stock-breeder of that name) was particularly outstanding in Affleck's opinion; and this Durham won prizes so consistently at Southwestern fairs that other cattle owners eventually ceased to enter their bulls in competition with him.[31]

By 1843 fine cattle had become commonplace in the old Natchez District. Many planters of the region had followed Hall's example and were importing Durhams. Others like James Payne of Jefferson County were building their herds from breeding stock of the Patton strain, a line of thoroughbred animals that had been developed in Kentucky from cattle imported from England. Dr. Asa B. Metcalf and St. John Elliot were breeding Ayrshire dairy cattle, and their work made these thoroughbreds popular as milk producers with planters of Southwestern Mississippi. A growing number of cattle raisers, however, were following the practice of improving their common "neat" cattle by mating their heifers to good Durham bulls belonging to Affleck, Hall, and other fanciers of thoroughbred livestock. Indeed, many of them considered the halfbreed Durhams more satisfactory milk cows, draft oxen and beef cattle than the more delicate pure-blooded stock.[32]

Thoroughbred Durhams were brought into the Vicksburg trade area in considerable numbers between 1839 and 1841; but their heavy mortality rate soon demonstrated that they were not well adapted to climatic conditions there. When turned loose to forage for themselves in the canebreaks and woodlots of Warren, Hinds and Madison County plantations, they often sickened on the rough diet and hot weather of Central Mississippi. Yet such cotton growers in this area as were willing to feed and care for their stock in the manner of Kentucky stockbreeders did succeed in raising part and fullblooded Durhams without undue losses.

Colin S. Tarpley was unquestionably the most successful of all the Central Mississippi stockbreeders and dealers during the depression.[33] He sold thoroughbreds and crosses from his stock of cattle, horses, sheep and hogs to planters all over the state; and many of his animals were prizewinners at the state fairs held at Jackson during the early 1840s.[34]

James F. Watson, a cotton planter who lived near Edwards in Hinds County, was another member of the group that bred improved cattle in the Vicksburg-Jackson area during the early 'forties. Watson managed to avoid the ruinous mortality rate common to thoroughbred Durhams in his section of the state by feeding his cattle and by servicing his cows with carefully selected bulls possessing no more than fifty percent Durham blood. Partblood sires of this kind produced "fat, sleek calves" when bred to the common Creole cattle of the Old Southwest. One of Watson's best bulls during the depression was a large, powerful animal descended from a red Durham bull and a cow raised in the Shaker community near Harrodsburg, Kentucky. According to Martin W. Philips, who was acquainted with both animals, Watson's "Jerry" was superior to Thomas Hall's "George Beltzhoover," and would have been a better show animal if he had received the care that Hall gave his pampered champion. Philips, however, would have been the first to admit that Watson did not neglect his stock. On another occasion, Philips described a splendid clover pasture that he had seen on a visit to Watson's plantation, "Bachelor's Retreat," at a date when both clover and cultivated pastures were exceedingly rare in Mississippi. In fact, Watson's success in breeding cattle was probably due primarily to the provisions that he had made for their care and feeding.[35]

Martin W. Philips himself enjoyed at least partial success in raising and selling improved cattle. His career as a cattleman and dealer began in January, 1840, when he bought a shipment of eighteen head of partblood Durhams from C. J. Blackburn of Kentucky.[36] Many of these cattle became ill when they were turned out to shift for themselves along the Big Black River during the late spring, and seven died from unexplained causes. This heavy loss, however, had been partly offset by the birth of several healthy calves and by the sale of a bull and a cow at a very good price. As no more of the Durham calves or cows died during the remainder of 1840, the survivors probably had become satisfactorily acclimated to their new diet and environment.[37]

When some of the sick Durhams recovered quickly after being turned into a field of green oats, Philips concluded that artificial pastures were the secret to successful livestock raising in Central Mississippi.[38] From that time onward, grasses and grains were standard crops on the small plantation beside the Big Black. Within a few years Philips's cattle, mules, sheep, horses, and pigs, fed adequately throughout the year, were regarded by competent observers as the best livestock in his part of the state.[39]

Although Philips never discontinued raising blooded cattle while he was master of Log Hall, his stock after 1843 was intended mainly for use upon his own plantation rather than for sale. Lack of adequate pastures and the excessively high initial cost of thoroughbred cattle restricted the market for fine livestock in the vicinity of Edwards. Furthermore, heavy losses suffered by pioneer Durham owners tended to deter even the wealthier planters from investing precious capital in the uncertain business of raising blooded cattle.[40] Consequently, Philips perceived before the middle of the decade that thoroughbred Berkshire hogs were safer and surer sources of income than Durham, Ayrshire, or Devon cattle.

After 1843 planters in all parts of the state gradually lost interest in Durhams. Although most cattle owners appreciated their merits, they were unwilling to give them the care and feeding they required. Moreover, many persons who had tried Durhams and Ayrshires thought that common Creoles, if properly fed, would produce equally good milk and beef. John C. Jenkins, Sr., a widely read authority on Southern fruits in the 1850s, for example, raised both Durhams and Ayrshires on his Adams County plantation, but he eventually decided in favor of the common "neat" cattle of the Natchez region. He admitted that many cattle owners "from want of care in breeding have allowed their native stock to degenerate." Jenkins insisted, nevertheless, that fresh importations of bulls from Texas and Mexico, "the source of our native cattle," would quickly restore the lost properties of their range animals if the principles of selective breeding were applied.[41]

Mississippians offered an additional objection to thoroughbred cattle on the ground that they were unsatisfactory draft animals. According to Thomas Affleck, the common "large brick-colored or brown oxen, with their singularly twisted ram-like horns," made excellent ox teams, while "Herefords, Durhams and Ayrshires . . . have done little good; the Durham least of all."[42] Jenkins echoed Affleck's opinion. "The Durham Breed," he wrote in 1854, "and I may safely say all the varieties of cattle from northern latitudes, are totally worthless as draft animals. . . . The native stock far surpass, in speed, in ability to endure heat, and in strength, any variety of northern origin."[43] Inasmuch as oxen pulled the heavier types of turning plows, transported the harvested crops overland to river ports or railheads, hauled cordwood and timber from the swamps, and performed the other heavy duties on ante-bellum plantations that would be done with tractors today, cotton growers were inevitably concerned with the quality of their ox teams. When it became known generally that expensive imported animals could not perform this necessary work satisfactorily, many cattle owners concurred with Jenkins that it had been "a sad mistake when many of our planters introduced from Europe and the North the famous Durham breed of cattle and discarded their native stock."[44]

This common prejudice against pureblooded Durhams, Devons, Ayrshires, and Herefords did not go so far as to include crosses between thoroughbreds and the native stock of Spanish descent. Instead, most Southwesterners believed that mixed breeds produced excellent beef and draft animals possessing some of the better characteristics of all their parent stock.[45] During the latter years of the depression, therefore, the trend in cattle raising was away from expensive and delicate thoroughbreds and toward the cheap and hardy crossbred range stock. Even so, pedigreed Durhams, Devons, and Ayrshires were never completely eliminated from Mississippi herds. A few enthusiasts continued to raise thoroughbred cattle, and almost all of the owners of large herds continued to service their cows with full-blooded bulls in order to preserve or build up the quality of their beef and dairy animals.

Although cotton planters of ante-bellum Mississippi always regarded their herds of cattle and hogs as more valuable than their flocks of sheep, they did come to hold sheep in greater esteem after 1837 than before. Prior to the depression, sheep had been raised in the Old Southwest almost solely for their mutton, and their number had been small. In fact, before 1839 only two flocks appear to have been maintained in Mississippi for the purposes of wool production, one of them belonging to B. L. C. Wailes of the town of Washington and the other to Mark R. Cockrill, of Nashville, Tennessee, and Madison County, Mississippi. Of the two flocks, only Cockrill's could have been considered an unqualified success. During the depression, however, wool production became a major consideration for Mississippi agriculturists. To conserve their scarce cash money, plantation owners installed hand-operated spinning and weaving equipment on the premises and detailed their least efficient farm hands to tasks of making thread and cloth for Negro clothing.

Early in the 1840s, home manufacturing of cotton and woolen textiles attained such dimensions in Southwestern Mississippi that John Robinson erected a steam factory at Natchez designed to supply local plantation looms with yarn and thread made from cotton and wool. Planters of the vicinity brought their cotton and wool to the Natchez mill to be spun into yarn, and, if they wished, their raw materials could be worked into finished cloth on steam-powered looms. Customers seldom paid cash for the services of the factory, but usually arranged barter deals with the management to make payment in raw cotton and wool. Whether planters patronized the mill or made their textiles at home, they all were anxious to supply their own raw materials; for they could raise sheep more cheaply than they could buy wool on the New Orleans market.[46]

Once they had begun to make clothing for their Negroes at home, slaveowners who had been interested formerly only in the quality of their mutton then undertook to improve their fleeces. Most of them began to build up the wool yield of the common Creole sheep by selective breeding and the use of thoroughbred rams. A minority, however, entirely discarded the native stock in favor of purebloods or crosses between different varieties of blooded sheep. By the latter part of the decade there were even a few stockmen—notably B. L. C. Wailes, Mark R. Cockrill, Martin W. Philips, Colin S. Tarpley, and Judge Alexander Covington of Warren County—who tried raising wool on a commercial basis. By far the greater number of cotton growers of the state, however, were content to shear only enough wool to meet the requirements of their own establishments; and very little wool was exported from Mississippi before the Civil War.[47]

During the depression, Mississippi sheep not only improved in quality but also multiplied in numbers.[48] Between 1839 and 1849 the flocks of the state increased from 128,367 head to 304,-929. This 133 percent increase was more than double that of any other domestic animal, an extraordinary expansion, due, of course, to the rising rate of domestic wool consumption.[49]

The story was quite otherwise in regard to the raising of horses and mules. Although planters made ever greater use of horse-drawn farm implements during the 1840s and 1850s most of them continued the practice of buying their mules and horses from Tennessee and Kentucky. For some undetermined reason, Mississippians were reluctant to purchase the stallions and jacks needed for home breeding purposes. Martin W. Philips in 1841, for example, was able to locate no more than four jacks in all Mississippi: one each in the counties of Hinds, Rankin, Warren and Oktibbeha.[50] And two years later populous and relatively prosperous Madison County was able to boast of no more than seven stallions and four jacks.[51] During the remainder of the depression, Mississippians bought both jacks and stallions in small numbers; but mule breeding never became common in Mississippi during the ante-bellum era. In the 1850s Mississippi mares foaled more colts than ever before; but home raised horses never filled more than a small part of the demand.[52]

Why horse and mule breeding never became important is far from clear. Farmers and planters continued to spend great amounts of money in importing horses and mules from states to the north; and they could scarcely have been indifferent, especially in the depression years, to the savings they could have effected by raising their own animals. The climate was favorable for raising colts, and no particular skill was required to bring them to maturity. On the other hand, grass of the types found in the stockbreeding districts of Tennessee and Kentucky was scarce and there were few cultivated pastures to fill the deficit. Here perhaps may lie the explanation.

It was evident during the 1850s that agricultural reformers had failed to persuade Mississippians to take up livestock breeding as an occupation separate from cotton growing. In spite of many cogent arguments in favor of cattle, sheep and hogs as sources of income, stock-breeding generally was undertaken only as an adjunct to the customary cotton routine. In fact, the change in Mississippi's livestock had been more qualitative than quantitative in character. During the 1840s, the number of hogs and horses and mules increased in almost exactly the same ratio as the slave population: the percentage growth of slaves was 59; horses and mules, 56; and hogs, 58. By comparison, the growth of the cattle herds of the state was very slight, only 18 percent. Yet in the cotton growing sections of Mississippi, growth of cattle numbers was larger than this figure alone suggests; for the cattle herds of the southern Piney Woods had been declining steadily. Even when this factor is taken into account, the fact remains that the number of beef cattle on the plantation was not expanding in proportion to increases in population in the same area. Only with sheep was the rate of growth greater than the expansion of the agricultural population.[53]

CHAPTER VI

Corn, the Principal Food Crop

The growing emphasis placed upon animal husbandry in Mississippi during the 'forties was accompanied by an equally important increase in quantity and variety of food crops. The panic of 1837 left Southwestern cotton growers no alternative but to grow their own livestock foods and breadstuffs for their families and slaves. The situation in Mississippi at the close of the 1837 cotton season was described thus by Colin Tarpley: "We . . . received our pay in worthless Bank paper of the country and paid it out and found ourselves in debt for the corn, pork and other necessary supplies that should have been raised upon our own farms. Hence we were driven from necessity to break our intolerable bondage to the grain growing states and raise within ourselves what was necessary for our own consumption."[1]

In America, corn has always been the basic food of farm animals and, to a lesser degree, of man himself; and Mississippians initially responded to the depression by planting more corn than had been their custom. Exactly how many additional acres they converted to corn cannot be ascertained now, because no reliable statistics on grains were compiled for the counties of Mississippi before 1840. Yet, the testimony of many witnesses indicated clearly that the output of ear corn and fodder, as dried corn stalks and leaves were called, was increased appreciably in all sections of the state where soils were suitable for agriculture. This change was under way even before cotton planters began in earnest to raise and cure their own supplies of pork.

109

Evidence of this trend was provided in one instance by an editor of the Natchez *Free Trader*.[2] After an extended tour of the central and northern counties of Rankin, Hinds, Madison, Yazoo, Holmes, Carroll, Yalobusha, Lafayette and Marshall, he reported in September, 1838, that he had observed large crops of corn growing upon every plantation along his route. At Grenada he had been surprised to see flour ground from Choctaw County wheat, and he had been told that wheat was currently being planted in many places in the Indian Cession country. Yet changes even more fundamental than those described by the Natchez editor were then under way in North Mississippi. Farmers in the neighborhood of Kosciusko were resorting to the hitherto unheard of expedient of reducing cotton acreages in order to increase production of corn and livestock.[3]

This determination to grow sufficient corn and fodder for home consumption was present also in the Southwestern cotton growing districts of the state. In June, 1838, the *Piney Woods Planter* declared that "Mississippi planters are generally turning their attention to raising such produce on their own plantations as their own necessities require." [4] And the New Orleans *Bulletin* explained in May, 1839, that Mississippi cotton growers had learned from the "recent derangements of the currency . . . to husband their resources by . . . growing their own corn and meat and all that their soil is capable of producing," instead of planting almost nothing but cotton as they had preferred to do in the recent past.[5]

Within three years after the beginning of the depression, corn was a major crop in the eyes of most Mississippians; and they were planting almost as many acres of it as cotton. This trend toward heavier corn production was especially evident in Central Mississippi. While reminiscing about agricultural developments in the Vicksburg trade area since his settlement there a decade earlier, a Hinds County planter remarked that very little corn had been produced in that area during the prosperous years of the 1830s, but he had witnessed great changes since 1837.[6] Where he once could have "travelled all day without seeing a corn field of any importance . . . now . . . beautiful fields of corn are everywhere to be seen." Almost every slaveowner that he knew was supplying his Negroes with food, and some of them were even selling small quantities to "non-producers of neighboring towns and villages." He concluded that his friends, who lately had prided themselves upon being planters and who cultivated nothing but cotton and imported "mouldy corn and tainted pork from up country," were now becoming more like Northern operators of diversified farms. A neighboring planter, Martin W. Philips, corroborated this report. In 1842 the master of Log Hall declared that rye, oats, wheat, turnips (for stock feed), clover and other grasses, and a surplus of sweet potatoes, pork and corn were then being raised in Hinds County.[7]

Further evidence of the growing self-sufficiency of Central Mississippi, at least in grains, can be found in the records of the railroad running between Vicksburg and Brandon. In 1850, for example, only 3,181 bushels of corn were shipped from the river-port into the interior, while more than forty-four hundred bales of cotton were transported in the opposite direction. Yet, in the same year, the railroad carried 297,119 pounds of meat from Vicksburg to Jackson and Brandon. Obviously farmers of Central Mississippi were doing much less to supply their own meat than to grow their own corn.[8]

The important place in Mississippi's agricultural economy that corn won for itself in 1838 was never challenged seriously during the remainder of the ante-bellum period. In fact, corn production was remarkably constant during the 'forties and 'fifties, whether times were good or bad. True, the crop did rise from thirteen million bushels in 1840 to twenty-two million bushels in 1850 and to twenty-nine million bushels in 1860. Yet the population of the state was increasing simultaneously at approximately the same rate. Hence, there was comparatively little fluctuation in the yield of corn: 35 bushels for each inhabitant in 1840, 37 bushels in 1850, and 38 bushels in 1860.[9]

When Mississippians took steps in the spring of 1838 to increase corn production, most of them began to plant one acre of corn for every two acres of cotton. This allocation of land must have provided adequate supplies of ear corn and fodder without undue waste. The two to one ratio was maintained virtually unchanged during the remainder of the depression, and was not modified materially even during the 'fifties when cotton again became profitable. On the other hand, there were wide variations from plantation to plantation in the average amount of land worked by a field hand and in the corn yield per acre.[10]

On a few Mississippi plantations, it became the rule to plant no more than three acres of corn and six of cotton for each farm worker. On others, it was the practice to require each worker to cultivate as many as six acres in corn and twelve in cotton. The great majority of cotton planters, of course, chose a mean somewhere between the two extremes.

The minority of planters who laid the lighter work loads on their laborers did so because they believed that to be most practical policy in the long run. One of them justified his theory in 1842: "We are cultivating less ground to the hand, with a view to better cultivation—preparing the ground more perfectly before planting the seed—paying more attention to the making and spreading of manure; to the ditching of our lands, and to rotation of crops." [11]

Farmers of this type often argued that cultivating small acreages thoroughly was more efficient and less expensive than working large crops of corn and cotton. They contended that their cotton fields were seldom "in the grass" as those of their fellows frequently were. Consequently, they had few crop failures, even in rainy growing seasons.

Some farmers subscribing to the rule of nine acres to the hand unquestionably were successful producers of cotton and corn, though perhaps not the most successful. For example, H. D. Bonney, of the Sartartia community of Yazoo County, owned enough Negroes to put into his fields a working force of twenty full hands.[12] Nevertheless he made a practice of planting no more than 180 acres in cotton and corn in the usual ratio of two to one. In a letter to the *Southern Planter* in the spring of 1842, Bonney reported that his operations of the previous year had been most satisfactory. After feeding and fattening 14,000 pounds of "Old Cornshire" pork, he had stored more than 4,000 bushels of corn; and his cotton crop had totalled 133 bales averaging more than 400 pounds each. His record of 67 bushels of corn, 700 pounds of pork, and 5¼ bales of cotton to the hand was worthy of respect by ante-bellum standards; and a cotton crop of almost 400 pounds of ginned cotton to the acre would not have been considered bad a century later. Bonney's was by no means an isolated example. With only ten full hands, Colonel H. D. Robertson, who lived near the town of Clinton, in 1842 gathered 100 heavy bales of cotton, 3000 bushels of corn, and 1500 bushels of sweet potatoes. Furthermore, Robertson fed three beef cattle in stalls along with seventy-five large hogs.[13]

Philips of Log Hall was in complete agreement with Bonney and Robertson that the goal of all good farmers should be economic self-sufficiency. Yet he had ideas that were diametrically opposed to those of Bonney and Robertson as to the means of achieving that end. Where they took pride in working small acreages intensively, Philips emphasized the use of draft animals and labor-saving devices and cultivated maximum crops of cotton and corn. The secret of success in extensive agriculture, according to Philips, was to depend upon horsepower instead of manpower. "I regard a good team and good implements as absolutely requisite upon any *farm* where the proprietor is anxious to make full crops," he wrote in 1849. Because inferior animals could not possibly perform the heavy tasks required, Philips went to great pains to buy top grade stock. These animals he kept in good condition by feeding them all the corn and fodder they would eat. Fearing the consequences of overworking horses and mules in the hot climate of Central Mississippi, he usually harnessed two horses to a plow that other planters would operate with one. According to Philips, the practice of using two horses instead of one prolonged the useful life of the animals, and permitted much deeper plowing. The latter consideration was important to both Philips and Thomas Affleck, for they were certain that deep plowing ensured bumper crops of corn and potatoes and at the same time preserved the fertility of the soil.[14]

On most Mississippi cotton plantations the tedious job of hoeing consumed a major part of the operative's time. Philips partially eliminated hoeing by plowing across the rows of cotton and corn and by cultivating the furrows with shallow-running horse-drawn cotton scrapers, harrows, cultivators, double shovels, and other farm equipment designed for specific tasks. The labor saved by these modern methods enabled him to plant crops that were unusually large by ante-bellum standards.

His custom was to lay out ten acres of cotton and eight of corn for each able bodied worker, plus small crops of miscellaneous foodstuffs such as oats, potatoes or timothy. In 1846, for example, there were 128 acres of cotton on Log Hall, seventy-two acres of corn, three acres of sweet potatoes, two acres of artichokes, three acres of millet, and twenty acres of rye and oats.[15] Since Philips's labor force then was the equivalent of eleven field hands, his average planting that year was twenty-one acres to the hand, an extraordinarily large average.[16] Philips took pride in being able to accomplish so much with so few Negroes, and he boasted that he could produce in an average season "six bales of cotton weighing 400 pounds, 200 bushels of corn, 500 pounds of pork, raise 5 lambs, 3 calves per hand that works in the field, and an abundance of oats, rye, millet and fodder, potatoes and garden vegetables, to feed one horse to each hand, all cattle in the winter daily, and whites and blacks." [17]

Philips's system of agriculture had one serious flaw, which Dr. E. Jenkins of Choctaw County took pleasure in exposing in letters to agricultural journals and local newspapers.[18] Jenkins admitted that large acreages of cotton and corn could be worked with horse-drawn farm equipment. But he insisted that they could not be harvested with horses and mules—cotton and corn still had to be gathered by hand. The Negroes would be compelled to work in the fields far into the months of winter if they were to gather all the cotton. Where, Jenkins asked, was the wisdom in preserving the health of horses and mules, while exposing more valuable Negroes to the cold winds of December, January and February? Philips never attempted to answer Jenkins's questions. Some, therefore, thought that the man from "the Horsepens" had found the chink in Philips's armor. Jenkins was not alone in accusing Philips of overworking his labor force. His immediate neighbors and even his own family regarded him as a harsh taskmaster.[19]

Between 1838 and 1849 production of corn per acre varied

greatly from region to region in Mississippi, and even from plantation to adjoining plantation. Differences in soil composition and soil fertility rather than in strains of corn accounted for most of these variations, but differences in individual farming methods also played a role of some importance.

The most abundant crops reported in Mississippi during the last two ante-bellum decades were grown on the incredibly rich alluvial soils of the Yazoo-Mississippi Delta, a swampy region that was still in the initial phases of settlement during the 1840s. Even without the aid of animal or vegetable manures, the deep black loam of the Delta produced bumper crops of corn and cotton. Yields of three thousand pounds of seed cotton to the acre were commonplace, and fifty to seventy-five bushels of corn to the acre were normal. Occasionally Delta corn crops exceeded even these high figures. In one instance, ninety bushels were gathered from a representative acre of land on a Washington County plantation belonging to W. P. Warfield, even though the field had received only ordinary care and no fertilizer of any kind.[20]

Corn was less prolific on the rolling hills of the northeastern, central and southwestern cotton districts of Mississippi. Fifty bushels to the acre sometimes were gathered under ideal conditions of soil and weather, but yields of as little as ten to fifteen bushels were not uncommon. In general, planters of the upland areas considered twenty-five bushels per acre a normal crop.[21]

The Piney Woods region of South Mississippi was generally unfit for agriculture, and its sandy soils would not support corn or cotton without manures or other fertilizers. In that part of the state, therefore, corn was seldom planted, except in garden plots where cattle had been penned the previous season, or in comparatively fertile "river bottoms" of the streams that flowed through the great pine forests.

Although white Mississippians had grown corn a century and a half before the panic of 1837, they still had no standard method of planting or cultivating it. Corn always had been so easy to grow in the Old Southwest that few farmers had ever bothered to determine its best method of culture. Indians, frontiersmen, and sometimes newly-arrived slaveowners, planted corn in natural clearings or between the trees in "deadenings" by dropping seed into holes opened with hoes or pointed sticks. Even under primitive modes of cultivation, this hardy grain had produced good yields. Where the spacing of the standing dead timber was great enough, or where fields were entirely cleared of trees, the farmers, of course, preferred for the sake of convenience to plant corn in drills and work it with hoe and plow. Beyond this point, however, there was little uniformity.

During the 'forties corn was planted in a number of ways. On Killona Plantation in North Mississippi, it was scattered in water furrows opened to depths of three or four inches and covered over with crude home-made cultivators possessing only two iron teeth. Rows on Killona were five and a half feet apart, and the hills in the rows were separated by intervals of two to two and a half feet.[22] Another system was used on a typical plantation in Hinds County.[23] The spacing of the rows, like that on Killona, was five and a half feet, with hills set at two-foot distances. The seed, however, were dropped in a trench and then buried with a turning plow. In a third method of planting, the seed were dropped in a drill on top of a raised seed bed, and then covered over with a harrow or drag in exactly the same method used for cotton.

While opinions differed as to the proper spacing of hills or rows, there nevertheless was general agreement upon the basic principle. All farmers believed that corn should be planted more thinly on poor land than on fertile soil. The rule, according to a correspondent of the *Southern Farmer*, was to "give just opposite treatment to corn and cotton as to poor land and rich—give corn *distance* (as we call it) only on poor land, but give cotton distance only on rich." In practice this particular farmer planted corn on his better lands as thickly as possible, leaving only space enough to plow between the rows. On less fertile fields his rows were spaced four or more feet apart, and the individual hills were separated by at least twenty inches.[24]

In 1839 Martin W. Philips became curious as to why every one in his acquaintance was planting corn in five foot rows with two foot intervals between stalks. He inquired among his friends but could find no valid reason for this practice. Apparently they were planting corn rows five feet apart merely because they spaced cotton rows five feet apart; and they were following the latter practice mainly because mature cotton plants were bushy and commonly attained diameters of three or four feet. Because nobody had ever experimented by varying the number of corn stalks to the acre, no one could tell Philips what would happen if he should narrow the space between the rows. As he was then obtaining unsatisfactory yields, he decided to study the question himself. In 1840, Philips brought his rows somewhat closer together than five feet. When his corn production rose as a result, he narrowed the gap further. In 1843 Philips planted his rows still closer; and the benefits obtained were impressive. With five foot rows he had been able to average no better than fifteen bushels to the acre on hill lands, even with the use of cotton seed fertilizer. But halving the distance between rows doubled that yield in 1843—the average that year was thirty bushels to the acre. So Philips made permanent his practice of planting corn more thickly.[25]

A. K. Montgomery, a friend and neighbor of Philips and his kinsman by marriage, was farming exceptionally fertile land along the Big Black River during the 1840s. On that land he used no fertilizer, spaced his rows five feet apart, and customarily harvested more than fifty bushels of corn to the acre. He was so much impressed by the results of Philips's denser planting and fertilization that he decided in 1844 to experiment with those methods on his own rich bottom lands. At Philips's suggestion, Montgomery scattered cotton seed broadcast over several acres at the rate of about five hundred bushels per acre and then turned them under with turning plows. Then the land was bedded up in the usual manner, and corn was planted in water furrows two and a half feet apart. When it came up, it was thinned with the hoe to intervals of one foot, instead of the customary two feet, and then worked during the growing season in the same way that he had always employed. At the end of the crop year Montgomery harvested ninety-eight bushels per acre from the experimental plot. Yet corn that had been planted in five foot rows without cotton seed fertilizer in an adjoining field yielded only forty bushels to the acre. To Montgomery the test was conclusive, and he promptly adopted Philips's system of corn culture.[26]

The spectacular results obtained by Montgomery and Philips inspired many of their neighbors to use more manure and cotton seed fertilizer on corn crops and to plant more thickly than had been their habit. By 1847 so many of them had accepted these changes that Philips reported to the *Southern Cultivator* that cotton growers of Central Mississippi had shifted from planting corn at the rate of "five by two (or 4346 stalks per acre) to four and a half by two, with two stalks in a hill—or even to a double row of four feet by one: that is, four feet from center to center." [27]

During the 'forties and 'fifties there was sharper disagreement among Mississippi planters over the best methods of cultivating growing corn than over questions of planting. Agricultural and political journals printed numerous letters advocating one or the other of two common methods of cultivating corn after the initial stage of growth was completed. The adherents of one school of thought argued that corn should simply be kept clear of grass by a harrow or cultivator and the roots left as much undisturbed as possible. The other group was equally certain that growing corn should be plowed as deeply as could be done with turning plows. In their view, deep plowing aided the penetration of water to the lower roots of the plants. Furthermore, it had the advantage, or so they thought, of cutting the lateral surface runners and thus forcing the plant to send out longer tap roots.[28] These differences over the relative merits of deep and shallow plowing raised such vigorous controversy during the early 1840s that Nathaniel G. North of the *Southwestern Farmer* remarked with a note of wryness: "scarcely an agricultural paper comes to hand but we see arguments on each side of the question, from different individuals who claim to have followed each plan with success. . . ." [29]

James S. Johnson, a planter of Jefferson and Warren Counties and the first president of the Woodville-St. Francisville railroad, was typical of those who believed in cultivating growing corn with a turning plow. His practice was to lay out six foot rows (because his land had been impoverished by forty years of cultivation in cotton and corn) and to plant corn in hills eighteen inches apart. When the corn was well out of the ground, Johnson had the dirt near the plants broken up to a depth of ten inches with a bull tongue plow. Negroes with hoes then scraped grass away from the remaining narrow strip of unbroken earth at the roots of the plants. While still small, the corn was worked over twice in this way. When it reached a height of two feet, it began to receive different treatments. Plows equipped with moldboards threw dirt over the roots; and turning plows broke the soil in the furrows, said Johnson, as deeply as "sharp ploughs and good teams will enable me to do." This deep plowing was repeated at intervals of one to two weeks until the green stalks had reached their full growth and the crop could be "laid by." Johnson was completely satisfied with his method of corn culture, believing that it produced as large a harvest as could be expected from his poor land.[30]

The shallow system of corn cultivation was described in some detail in 1841 by a correspondent of the Jackson *Southron,* writing under the pen name of "Tuscarora." In contrast to Johnson's practice of deep plowing with bull tongues and turning plows, "Tuscarora" preferred to pulverize the earth and destroy grass and weeds with harrows, cultivators, and the slightly deeper-running double shovels. His only deep plowing was done early in the season while the land was being prepared for planting. At that time turning plows, following one behind another, were used to rip open the soil to depths of six or eight inches, burying grass, stubble and cornstalks in the process so that they might "retain moisture and decompose." After the corn had been planted (in a water furrow) no more subsurface plowing was done, lest the roots of the growing plants be disturbed. Instead the ground was loosened by repeated workings with cultivators and double shovels. Toward the end of the cultivation period, "Tuscarora" sowed peas broadcast between the rows and covered them with harrows. At ten foot intervals in every fourth row he planted pumpkins as winter feed for hogs and cattle; and an occasional row of Irish potatoes served the needs of his family and slaves.[31]

The persistent argument over the relative merits of shallow culture and deep plowing was never resolved in the ante-bellum era, with both systems continuing in use beyond the Civil War. Even so, the more modern shallow culture seems to have been making headway against the other method of deep plowing. Beyond a doubt, implements designed to pulverize the surface of the earth were gaining in numbers, types and popularity during the 1840s and 1850s.

Aside from Martin W. Philips and a few other scientifically-inclined cotton planters, not many Mississippians were interested in improving their varieties of corn or their methods of corn cultivation. Several planters were actively engaged in developing new and better strains of cotton by cultivation of accidental mutations or by quasi-scientific selective breeding techniques; yet almost none bothered to extend his activities from cotton to corn. Nor were large quantities of seed corn bought from Northern professional corn breeders during the 'forties. Instead, year after year, most Southwesterners were content to plant the same varieties that had been raised in the the Lower South for generations—strains that their forebears had obtained originally from Indians or brought into the state from the East.

The only distinctly different varieties of corn planted in significant quantities in Mississippi during the 1840s were Flint and Gourd Seed. For a number of reasons the former was the more popular with Southwestern farmers. Crosses between the two, however, were very common. The hard, rock-like grains of the Flint variety made it resistant to the attacks of weevils and other insects. Also, Flint's white meal was believed to be better for making bread than meal from the Gourd Seed variety. On the other hand, Gourd Seed was the more prolific of the two. Nevertheless, Flint's superior food qualities made it the favorite.[32]

Mississippi agriculturists of the 1837–60 period seldom bothered to change their strains of corn. Philips, for example, cultivated a variety that he called Two-Thirds Flint, the seed of which he obtained from a planter who had raised it for forty years.[33] Thomas Hall customarily depended upon a Gourd Seed derivative called "Turcarora," purchased from John W. Brisson of Washington, Mississippi. Brisson claimed to have developed Tuscarora by careful selection of ordinary Gourd Seed, preserving it without "Change or inter-mixture" for some thirty years.[34]

One who did not conform to the common practice of planting

either Gourd Seed or Flint corn was Dr. Benjamin F. Johnson of Hillsdale in Warren County. He was a professional corn-breeder, a planter of cotton, and a dealer in seed as well. In 1845 Johnson obtained seed corn from the Allegheny Mountains and cultivated it three years on his Warren County plantation. Each year he carefully selected seed for the next crop from the largest ears and stalks in the field. Within a short time the mountain corn showed amazing improvement. In 1848 a four-teen-acre plot of Johnson's Allegheny Golden Grain yielded 1525 bushels, a record crop of 109 bushels to the acre. Johnson's corn was not only prolific, but heavy: an average bushel of the Golden Grain weighed 60 pounds while ordinary corn averaged only 52. The extraordinary characteristics of Golden Grain encouraged Johnson to place its seed on the market. In 1848 half bushel lots sold at ten dollars in Jackson, Yazoo City, Vicksburg, and Natchez. Unfortunately, no further evidence about Johnson's venture into corn breeding has been uncovered. However, the fact that Allegheny Golden Grain failed to attract any further notice in the press, and that Flint and Gourd Seed remained the favorites in Mississippi, indicates that its popularity was never widely established.[35]

Mississippians usually selected corn with hard white grains rather than the yellow varieties when making corn breads and hominy. In grinding corn for the table, millers set their stones to crack open the flint-like cover of the grains without crushing them into powder. The resulting large hard particles then were sifted out of the meal and converted into "small hommony," or grits, which Negroes ate with molasses or buttermilk. The corn meal left after the bits of broken corn had been removed was reserved for hoe cakes, johnny cakes, or the many variant forms of corn bread produced in Southern kitchens.[36]

During the 1840s and 1850s, farmers continued their old practice of feeding corn to livestock, both in the ear and as fodder. When feeding horses, cattle and hogs, they usually threw the whole ear of corn into the troughs; for shelling corn by hand was a slow and tedious process. On more progressive plantations, however, mechanical corn shellers, grinders, and straw cutters were coming into use by the end of the 'forties. Owners of these machines usually maintained that corn and fodder became better food for cattle and horses when the grain was shelled from the cob and the dried leaves and stalks shredded into smaller particles.[37]

In the Old Southwest corn shucks also were fed to livestock. Thomas Affleck, for one, considered them to be more nutritious than fodder. In his opinion, both shucks and fodder were roughly equivalent to timothy hay pound for pound. However, Affleck regarded all three as inferior to Bermuda grass hay. Like a small minority of planters, the journalist believed that the common practice of obtaining fodder by stripping leaves from the growing plants was injurious, reducing the corn yield by as much as ten percent. Some men of this opinion followed the custom of planting corn fields to be harvested as fodder only. In such fields, corn was planted in drills and covered over with harrows or sowed broadcast and buried in the same way. The corn stalks were gathered just before the first ears appeared and then dried in the sun. Regardless of the method of harvesting, corn shucks and fodder were the principal winter livestock feeds.[38]

It is evident, in retrospect, that the evolution of corn culture had run its course by 1850. Such changes as occurred in the 'fifties were nothing more than continuations of trends established during the depression. In the main, farmers continued to plant the same varieties that they had raised in the 'forties. Few persons went to the trouble to import new strains from other states, and even fewer tried to improve their own plants by selection of seed or hybridization. The story was much the same as to methods of cultivation. No change took place in the basic principles of working and planting corn during the last ten years before the Civil War. Farming implements used in the corn field underwent improvement, manual labor was slowly replaced by the work of horses and mules, and increased attention was being given to building up the fertility of the soil with animal manures, cotton seed fertilizer, and legumes. All these developments, however, had originated during the depression decade. There were to be no further innovations until scientific hybridization was introduced into corn breeding after the Civil War.

CHAPTER VII

Minor Food Crops

Corn was by no means the only foodstuff affected by the 1837-49 depression in Mississippi. Cowpeas, oats, sweet potatoes and upland rice, which had been cultivated on limited scales before 1837, were stimulated by the almost universal desire of Southwestern planters to "live within themselves." A large assortment of new grasses and grains was introduced with varying success. Wheat, rye, barley, millet and peanuts were planted in appreciable amounts for the first time, especially in the central and northern cotton growing districts. At the same time fruit orchards and cultivated pastures of Bermuda grass, clover, timothy, alfalfa (then known as lucerne), Kentucky blue grass, Texas mesquite grass, herds grass, orchard grass, and many others with quaint, old-fashioned or exotic names, rapidly lost their status as curiosities.

During the ante-bellum era none of these food crops, with the exception of fruit and possibly wheat, was ever produced in Mississippi for sale in out-of-state markets. Yet, despite their strictly local use, these grains, grasses, roots and legumes became important. Their addition to the traditional Southern crops of cotton and corn in the 'forties and 'fifties gave to Mississippi the nearest equivalent to a diversified and self-sufficient agriculture that the state was to possess until the 1929-1940 depression and the Second World War wrought a near agricultural revolution.

Because federal censuses conducted before 1849 all ignored cowpeas, the absence of statistics on that legume has tended to conceal its value to Southerners. The slightly distorted picture of Southwestern agriculture drawn by scholars who depended upon these censuses could have been corrected by the use of information readily available in contemporary newspapers and agricultural periodicals. Countless letters written by farmers and cotton planters all attested that cowpeas were a major source of food for cattle, horses, mules and hogs in all cotton and corn growing regions. These letter writers allowed their livestock to graze on the ripe peavines in the fields, and then plowed under a part of the vines to improve the fertility of the soil. In this section, where Northern horse or bull rakes were almost unknown, peavines were almost never cut and stored as hay. Only the ripe peas were gathered as seed for planting. The fact that Southwestern planters generally harvested nothing but planting seed affords a clue to the total size of the cowpea crop. This crop of seed in 1850 amounted to 1,072,757 bushels. Mississippians planted from a third to one bushel of cowpeas to the acre. The 1850 harvest, therefore, could have sowed more than a third of the 3,444,358 acres of land under cultivation in Mississippi. Because corn occupied about a third of the improved land in the state, it is possible—and even probable—that cowpeas were being planted in virtually all corn fields. In any case, peas seem to have occupied about as much land as corn—even though the fields used for both were generally the same.[1]

Thus the agricultural economy of ante-bellum Mississippi and her Southwestern sister states rested solidly upon a triple foundation of cotton, corn and peas. The widespread use of a legume to enrich the soil also proves conclusively that cotton growers of the 'forties and 'fifties, whether small farmers or great slaveowners, deserve more credit as skillful agriculturists than has since been accorded them.

While there is some question as to exactly how many Mississippians followed the practice of sowing cowpeas in their corn fields, a great majority undoubtedly did plant them for use as stock feed and vegetable fertilizer. True, many of them had only vague or incorrect notions about why peavines when plowed under as "green manure" improved the fertility of the

fields. Yet they could have learned the correct explanation from
the agricultural press. R. L. Allen, of the very influential *American Agriculturist*, knew about the nitrogen-extracting abilities of
clover, cowpeas, and other legumes, and made this information
public in 1849. Writing on various properties of the Southern
field pea, Allen stated unequivocally that it was the "best of
fertilizers for the South." "The cow pea," he explained, "is an
economical fertilizer, in consequence of its broad, succulent,
bean-like leaves, drawing nitrogen and carbonic acid largely
from the air. . . ." [2]

That peas restored the fertility of soils exhausted by continuous cropping in cotton and corn was a fact well known to
Mississippians as early as 1842. In June of that year, A. G.
Ailsworth of Madison County wrote to the editor of the *Southwestern Farmer* that every cotton planter would be wise to.
follow all of his corn crops with peas.[3] Not only were peavines
valuable food for stock during the months of autumn, he stated,
but they were also "amply worth cultivating for their renovating
qualities to the soil. . . ." Peas grown upon corn land, according
to Ailsworth, had the effect of restoring more fertility than corn
had extracted, provided that the peavines were plowed deeply
into the ground.

Ailsworth's belief in the virtues of peas was held by John B.
Stanley, of Utica, Mississippi, and many others.[4] Years of
experience in raising peas, said Stanley in 1849, had convinced
him that they would renew the productivity of land so poor that
nothing would grow on it but weeds and sedge grass. "No
operation to which land can be subjected," he added, "puts it in
better state for a succeeding crop than when a heavy crop of
vines has just been taken off of it . . ." In the same vein Goodloe
W. Buford of College Hill, in Lafayette County, declared:

> The cornfield pea—why, sir, it is the best renovator
> of the soil we have in this part of North Mississippi;
> and it not only reclaims, but prevents our land from
> wearing out by shading and sheltering it from the
> burning sun of July and August and the drenching and

washing rains that sometimes fall in the summer and
autumn, while it also supplies a coat of litter that
nothing else will furnish to the land, and which it
greatly needs.[5]

More concisely E. Jenkins, of "the Horsepens" in Choctaw
County, summarized the role of this crop when he stated that
"Peas are . . . the Clover of the South." [6]

Field peas were not confined wholly to corn fields in the
depression era. One correspondent of the *Southwestern Farmer*
who was sufficiently enthusiastic about them to write under the
pseudonym "Pea Vines," habitually sowed peas in the stubble
of oat and rye fields, as well as on lands that had first been
planted in corn.[7] When the vines ripened in his cut-over oat and
rye fields, "Pea Vines" had them mowed with scythes, dried in
the sun, and then stacked in cocks for cattle food during the
winter months, a practice that was relatively uncommon. Although Martin W. Philips also liked to grow cowpeas on land
that had just produced an oat crop, he grazed his cattle upon
the peavines in the field instead of cutting them for hay, and
then turned the residue under with plows in order to enrich his
soil.[8]

While many Mississippi farmers and planters were certain
that common cowpeas, next to cotton seed, were their most
valuable fertilizer, even more of them were accustomed to
depend upon peavines for stock feed during the fall when
natural grass was scarce.

None in the early 1840s was more in favor of peavine fodder
than Martin W. Philips, who went so far in 1842 as to state in
a letter to the *Southwestern Farmer* that "they pay better than
any other crop, as the amount of labor is not at all to be considered with the advantages resulting . . . they fatten everything
and have no expense in gathering, for cattle, hogs, sheep, and
horses will gather them when they have nothing else to do." [9]
Philips subsequently had the misfortune to lose valuable cattle
and hogs at a time when they were feeding in his peafields, and
he concluded that rotting peas and vines were dangerous to all
livestock.[10] When this opinion was expressed to the press, he

stirred up a controversy which continued intermittently throughout the remainder of the ante-bellum period, and attracted letters from partisans in all parts of the Lower South.[11]

The tone and distribution of these letters made it clear that a majority of Southern livestock owners were allowing their herds and flocks to graze in peafields without injury to the animals. A comment of E. Jenkins was typical: "I have been raising peas for stock about twenty-three years . . . and two-thirds of the time . . . I have fattened my meat on them without any other kind of food, with the exception of scattering pumpkins and corn occasionally over the fields; and they eat them in every stage—green, frost-bitten, dry, wet and sprouted."[12] Eventually Philips himself reverted to the practice of feeding cattle, sheep, horses, mules and hogs upon peavines in the field. He always was careful, however, to remove them from the peafields when the vines began to decompose.

Planters in Mississippi had been familiar with the value of sun-dried peavine hay long before the panic of 1837; and a few of them harvested peavines as hay. "It is an important means of keeping stock," a Copiah cotton grower wrote in 1842, "which, if well cured, will do better than any other species of forage we have."[13] Most Southwesterners, however, made no effort to gather peavine hay in quantity because of the great difficulty encountered in mowing and gathering masses of vines from the fields where they were entangled with corn stalks and weeds.

Hopeful of overcoming this troublesome obstacle, John B. Stanley devoted several years of labor to developing a horse-drawn machine which could rake vines and bundle them together in long rolls that could be handled easily.[14] His device was patented in 1849, but it failed to give satisfaction to potential customers. As no one else was able to succeed where Stanley had failed, peavine hay continued to be a rarity in Mississippi until after the Civil War. During these years most farmers and planters continued to feed their stock on peavines in the fall and, during the winter and early spring, on dried cotton seed, ear corn and fodder.

Production of sweet potatoes for hog feed and human use increased during the 1840s to an even greater extent than the output of corn or fieldpeas. Between 1840 and 1850 the Mississippi potato crop rose from 1,630,100 to 4,741,795 bushels, a jump in annual *per capita* production from 4.3 to 7.8 bushels.[15]

In this period, as later, sweet potato yields were large, and a little land was sufficient to provide all the potatoes needed by farm workers. In 1842, for instance, one cotton grower reported to the Washington (Mississippi) *Southern Planter* that he had gathered enough yams from twenty hills to feed his slaves and the members of his family for an entire year, and the potatoes left in the ground in his twelve acre plot had fattened eighty-five hogs (which had produced 10,000 pounds of pork when butchered).[16] In Yazoo County, in the same year, another farmer had dug ninety-seven bushels of yams, some weighing six to seven pounds each, from one quarter-acre field.[17]

Growing sweet potatoes was never complicated. By 1840, several generations had simplified production and labor so far that no important changes were thought necessary. In planting, pieces cut from a yam were set out in beds similar to those thrown up with the turning plow for cotton, with distances of six to eight inches between the "hills." After the potatoes had begun to sprout, plows were run down the furrows to kill grass, and hoes were used to clean the small spaces between the hills. All cultivation was discontinued as soon as the vines began to spread from the rows into the middles, and the potatoes were left to grow as they would until ready to gather. When mature, the potatoes needed for human consumption were dug up and stored in bins under a roof in order to prevent rotting and sprouting. The remainder of the crop was left in the ground for hogs to forage on. Upon occasion sweet potato vines were gathered for fodder or grazed in the field by cattle, and frequently they were plowed under as "green manure."[18]

Although the sweet potato was the only root crop planted in quantity in Mississippi during the 1840s, in that decade two others were introduced by farmers who read agricultural periodicals. The turnip was one of these. A few planters grew it with moderate success as feed for livestock. Yet turnips did not become popular with stockbreeders, because sweet potatoes already filled the need. On the other hand, a second crop brought into Mississippi in the 'forties had real potentialities for the future. Peanuts, or "pindars," as the Negroes called them, were raised in ever-increasing quantities during the 'forties and 'fifties as feed for swine. The trend toward large scale production gathered momentum slowly, however, and peanuts did not attain importance until the closing years of the ante-bellum period. By that time, they had been accepted at least by the better educated class of agriculturists as a valuable source of food for both man and beast.[19]

The production of upland rice, which had been cultivated in the southern half of the state for many years, spurted upward during the 1840-50 period, rising from 777,195 pounds in 1840 to 2,719,865 pounds ten years later. The upland variety of rice was the only grain known to grow well on the sterile sandy soils of the Piney Woods region; and it was planted principally as a source of stock feed by planters and cattlemen of the pine barrens and the Gulf Coast. Its culture, like that of most small grains, was simple. The seed were planted in drills between the growing pine trees on the sandy hills typical of that section of the country. The plants received no cultivation beyond a plowing or two while they were still small. There was very little undergrowth and only sparse grass to contend with in the pine woods. Hence hoeing was unnecessary under ordinary circumstances. After gathering, the grain could be stored indefinitely if not attacked by weevils; a characteristic which made rice a good stock feed for winter and early spring.

Upland rice was often served on the tables of South Mississippians; yet little of it was sent to New Orleans where there was a market for it. An almost total lack of cleaning machinery and equipment accounted for this failure to raise rice for sale. During the 1850s, a steady decline in the South Mississippi cattle industry reduced the demand for stock feed in that locality, and this development in turn caused a falling off in production of upland rice. By the close of the ante-bellum period, the annual crop had dropped from nearly three million pounds to a level of only 800,000 pounds.[20]

In the pre-depression era, breads made from wheat had been

luxuries unavailable to slaves and poorer whites in Mississippi, and only comparatively prosperous cotton growers had been able to afford "light breads" regularly. Because of a common belief that wheat could not be raised successfully in the Lower South, none had been planted in Mississippi. All flour consumed in the state had been imported from the Upper Mississippi Valley. After 1837, however, most cotton growers in the Southwest found themselves with very little cash. Hence the stream of flour that had poured over the floating wharves of Natchez and Vicksburg during the boom years of the early 'thirties was quickly reduced to a trickle.

People who had been accustomed to eating wheat breads were understandably reluctant to subsist entirely upon corn-bread, and in the fall of 1837 a considerable number began to experiment with wheat. When tested, the old theory that wheat was not suited to conditions in Mississippi proved invalid. James McGehee of Zion Hill, for instance, reaped forty bushels of good wheat in the spring of 1838 from only a bushel and a half of seed. Others obtained similarly encouraging results, and the news quickly spread that wheat could be grown on the plantations of Central and North Mississippi. People in the southern half of the state, however, were not so fortunate. Their experiments often ended in total or partial crop failures: the climate of the Natchez area was simply too warm for the successful cultivation of wheat.[21]

By 1839 wheat had become a crop of some consequence on many farms and plantations of North Mississippi.[22] In July of that year the Holly Springs *Banner* reported: "Marshall County has raised as much wheat this year as she will need, and it is a source of proud gratification to us that we are and will continue to be independent in future of our Sister States for bread stuffs." [23] According to the Columbus *Democrat,* a similar situation existed in Lowndes County.[24] Other northern counties experienced the same trend. By 1840 Yalobusha County was raising 15,000 bushels annually; Lowndes, 14,000; Tippah, 15,000; Lafayette, 9,000; Marshall, 19,000; and Monroe, 23,000. The total harvest for the northern half of the state that year was 156,820 bushels.[25]

After its introduction, wheat culture gradually moved south-ward over Mississippi, until it eventually was being cultivated successfully, along with cotton, corn, cowpeas and sweet pota-toes, as far south as Kemper, Hinds, Rankin and Madison counties.[26] In November, 1842, an editor of the DeKalb *Olive Branch* stated categorically that "the idea that Wheat could not be grown to advantage in this country is now completely exploded." [27] As evidence, he cited several abundant harvests of wheat grown in Kemper County, which in quality compared favorably with Northern wheat. Areas west of Kemper also pro-duced good crops. In 1842 D. O. Williams of Hinds County gathered his third successful harvest of wheat, and his neighbors were so impressed that they bought his entire crop to use as planting seed.[28]

In the 1840s wheat production expanded so rapidly in Mis-sissippi that bolting cloths for sifting flour were attached to most of the commercial corn, or grist mills in the state; and several true flour mills were erected at strategic points.[29] One was established in Vicksburg in 1843 by the firm of Vannetta and Fulsome.[30] The most important flour mill in ante-bellum Missis-sippi, however, was set up in conjunction with a steam-powered cotton and woolen textile factory at Bankston in Choctaw County. There, in the late 1840s, the new Mississippi Manufac-turing Company was finding times hard and profits meager. As a result its resourceful president, James M. Wesson, began to cast about for means of increasing the company's income. Having ample steam power, he decided to install mills for grinding corn and milling flour. Some wheat was already being grown in the vicinity of Bankston, but not enough to keep the mill in steady operation. Wesson, therefore, decided to persuade farmers to plant larger crops of wheat. He not only lectured and wrote about the advantages of grain growing for Choctaw and the neighboring counties, but also went to the trouble and expense of importing seed wheat from out of the state for distribution among interested farmers. Wesson's educational program met with extraordinary success, and by 1857 the mill was grinding local wheat at the rate of 60,000 bushels a year.[31]

Wesson and other reformers were fond of saying that wheat

culture was adapted to the requirements of cotton farms and plantations of all sizes. They pointed out that the grain was sown in the late fall when labor could be spared from the cotton fields; and once planted, it needed no further cultivation. Furthermore, it matured in May or June, during a comparatively slack period in the cotton growing season, when reaping and threshing interfered but little with cultivation of cotton and corn. Wheat yields in Mississippi varied from fifteen to fifty bushels to the acre according to the season and the richness of the soil. This rate was similar to that of corn grown on comparable lands. Yet it was achieved with infinitely less labor than was required for corn. Because of its small labor requirements, wheat was particularly well suited to the needs of small cotton farms where there were few Negroes, or none at all. Therefore, it became an integral part of agriculture in North Mississippi, where large plantations were mingled with numerous small farms.

Fairly representative of the slaveless farmers who grew wheat in the 1840s was Ferdinand L. Steel, who owned a small farm near Grenada. All of the labor on the Steel farm was performed by Ferdinand, his brother, his sister and his widowed mother. Cotton, corn and cowpeas were their principal crops; but Steel also planted oats for his horses and cattle, and a few acres of wheat. The Steels possessed very little farm equipment, at times lacking even a cart. They were obliged, therefore, to gin their cotton and grind their grain on shares at plantations of wealthier neighbors. They sold a little corn from time to time, but no wheat, for they consumed all their flour at home. Steel himself disliked the growing of cotton, but was unable to find any substitute for it that yielded as much money for the effort expended. Therefore, he and farmers like him continued during the 'forties to depend principally upon cotton for cash income, while raising their own supplies of corn, wheat, vegetables and meat.[32]

A few years after wheat was introduced into Mississippi, it began to be plagued with fungus diseases. These fungi soon grew into a serious threat, causing such widespread damage that they helped to prevent wheat from replacing cotton as the principal market crop in the northern tier of counties. To check the ravages of "rust" (as the fungi were collectively called), farmers tried various expedients. Some scalded their planting seed in boiling water immediately before sowing; some soaked the seeds in a weak solution of lye or "Blue Stone" (blue vitriol); some tried to buy their seed from stock known to be free of infection. By these means planters in the northern half of the state raised enough sound wheat to meet their own needs and still have a surplus to sell to the wheatless southern portions of Mississippi. Consequently, the farming community of the state was able to remain comparatively self sufficient in the production of wheat during most of the 1840-60 period.[33]

Several other cereals were planted in the cotton-growing sections of Mississippi, but none was cultivated as a grain. Progressive livestock owners used oats, rye, barley and millet as pasture grasses. While the fields were still green, stock were turned into them to graze. As a rule the owners harvested only small quantities of ripe grain as planting seed for the next season. Federal statistics on these grains, therefore, provide no reliable information about the extent of their cultivation. Yet other evidence to be mentioned subsequently suggests that oats and perhaps rye were planted on a fairly large scale as supplements to the major forage crops of corn and peas.

Increasingly during the depression and afterward, Southwesterners were troubled by problems of feeding their livestock in seasons when native grasses were scant. In this era, land was constantly being cleared of timber and brought under the plow, a trend which steadily decreased the range available for pasturage. A further complication was the rapid disappearance of the great canebreaks which formerly had fed vast numbers of cattle and horses during the months of winter, but which now were in danger of extermination by overgrazing, brush fires, and deliberate clearing for "new ground." As natural sources of provender diminished, planters were forced to provide fodder during the winter and early spring. Feeding in summer and autumn was not so great a problem. Natural grasses in plantation wood lots still provided most of the grazing necessary during the late spring and summer; and peavines were enough to carry horses, cattle sheep and hogs through the months of the fall season. But winter feed was drawn largely from supplies gathered and stored during the summer and fall.

In ordinary seasons most agriculturists were able to put up enough corn fodder, ear corn, and cotton seed to feed their live-stock through the winter, but poor corn crops caused by unseason-able weather were not unusual. At such times, farmers who de-pended mainly upon ear corn and fodder for their winter feed were obliged to buy corn and hay from the Upper Mississippi Valley, and this outlay was particularly distasteful during the depression. The larger the herd, the more pressing was the owner's need to provide spring, fall, and winter pastures.[34]

The oat was a favorite pasture grass with Mississippians in the 'forties, and several varieties were planted in moderate quan-tities. Martin W. Philips once reported that he had seen oat fields in the late spring covering as many as two hundred acres on Central Mississippi plantations. These were exceptional, but small spring oat crops were common.[35]

The ordinary variety of oat proved unsatisfactory as an early spring pasture grass. It had to be planted in the late fall in order to provide grazing in March and April, and it proved so suscep-tible to cold that it could not withstand the occasional freeze of Mississippi winters. A strain known in the South as the Winter or Egyptian oat was introduced in the middle 'forties by several agricultural reformers. The Egyptian oat was found to be hardy enough to survive all but exceptionally cold Mississippi winters. At the instigation of Philips, Affleck, North, and other agricultural writers, many planters owning blooded live-stock began to use the Egyptian oat for spring pastures in the latter years of the depression; and some of them even adopted the practice of planting winter oats in cotton fields late in the fall to check erosion.[36]

Rye, barley and millet were planted much less widely than oats and were used entirely for spring and summer grazing. Experiments were conducted with rye early in the depression to determine its usefulness for spring pastures; but they were dis-continued when it was learned that rye was liable to be killed by the cold. Inasmuch as natural grass areas were at their best in late spring and early summer, there was less need for artificial grasses then than at any other period of the year. Hence crops of summer millet and barley were never common in ante-bellum Mississippi.[37]

During the 1840s some of the more progressive livestock owners decided to try to build up summer and fall pastures for their thoroughbred animals, even though the need for such pastures was less urgent than for grazing lands in winter and early spring. Consequently, they imported various grasses from several sources in Europe and the United States and planted them in experimental plots. Very few considered the results encouraging. In most cases the fault lay in the livestock owners rather than in the grass, for they were usually seeking an ideal plant that simply did not exist. What they wanted was a grass that would grow luxuriantly upon fields too exhausted to produce good crops of cotton and corn, one that required no cultivation after planting, and one that would withstand grazing many months of the year. Such men were obviously ignorant of the most basic fundamentals of pasture construction. One observer reported in 1849 that Mississippians almost always carried out their tests of grasses on scales much too small for conclusive results, and they were too impatient to wait for the grasses to establish themselves firmly in the field. "When the experiment is going well and promising much," he remarked, "all the stock is turned in on it, and it is overgrazed. . . . Because an acre, or perhaps two or three (five is always a very large one in Mississippi) will not survive all these impositions, the conclusion is at once; this grass will not do." [38]

Clover of the red and white varieties was the first pasture cover to be tried and discarded by Mississippi livestock men. During the 'twenties and 'thirties it had been widely cultivated in the North as a soil restorer and as green food for livestock, and its reputation had spread into both the Upper and the Lower South. About 1835 seeds of this well-known plant were imported from the North, and sown on the red clay hills of the old Natchez District. For reasons unknown at the time, every test

made during the late 1830s was a total failure. The experimenters were mystified. It was not till 1842 that Nathaniel G. North,
in an article published in the *Southwestern Farmer*, explained the
chief reason for the failure. He wrote:

> It has been asked perhaps a thousand times—Why
> will not Clover succeed in Mississippi: It has been tried
> perhaps by every farmer in the state who ever had a
> notion of stock raising; and although in a few places we
> have seen it succeed, yet generally it is well known that
> no manner of treatment it yet has received has pre-
> vented its totally failing. The true cause of this failure
> we think brought to light in Mr. Edmund Ruffin's report
> to the state board of agriculture in Virginia. He says
> that in that state it was almost impossible to raise
> Clover until the application of lime was adopted. . . .[39]

North's diagnosis was correct. When lime was added to the
soil, clover grew satisfactorily in the Natchez area, provided
that the soil was sufficiently rich. It also did equally well in
other parts of the state. Although a few planters, particularly
in the Vicksburg trade area, built up pastures of clover during
the depression, the number who were willing to remove cotton
and corn from fertile land in order to plant clover or other
grasses was small. Furthermore, they were even less willing to go
to the trouble and expense of spreading crushed marl or plaster
of Paris over their pasture lands. Hence red and white clover
fields remained scarce in Mississippi.[40]

Toward the close of the 1840s the attention of stockmen in
the cotton states was attracted by reports of a valuable pasture
cover then being raised in Mississippi, and at least a few Ala-
bamans and Georgians are known to have ordered seed from
Mississippi growers. The plant, which had been discovered in
Noxubee County by Elisha B. W. Kirksey in 1840, was a variety
of clover requiring no lime. Most Mississippians thought it indig-
eneous to the Southwest. According to a description published
in the *Mississippian* in 1849, the "Yellow Clover" was a vine having a pronounced yellow spot in the center of each leaf. The

vines ranged in length from one to eight feet, and when the plant was in bloom a small yellow blossom appeared at every leaf along the vine. The burrs were round and "put together in a spiral form like a snail's shell." Two crops of yellow clover a year could be produced in the latitude of Natchez, one in the spring and one in the fall. The spring crop went to seed toward the end of May and the plants subsequently died. In late summer they reappeared and assumed a luxuriant growth in the months of fall.[41]

When Kirksey first saw a patch of yellow clover growing wild along the banks of the Noxubee River, he was so impressed by its rank growth that he had the plot put under fence. In subsequent years he allowed his stock to feed upon this native yellow clover under carefully controlled conditions, and cattle, horses and swine all appeared to find it appetizing. After observing Kirksey's pasture of yellow clover for several years, a neighbor reported: "It proved to be superior to any pasturage I have ever seen for hogs, cattle, horses and goats, all of which I have known to live on it for a long time and were much benefitted." [42]

Yellow clover won the approval of planters in the vicinity of Natchez when they learned that it was suitable for winter grazing in their warm climate; and several of them developed yellow clover pastures in the closing years of the decade. When Solon Robinson visited the southern part of Mississippi in 1848, he saw fields of it, but was unable to identify the plant.[43] In 1849, however, an unidentified planter of Amite County learned that it was the same plant mentioned in *Gardiner's Farmer's Dictionary* as French Clover *(trifolium incarnatum)*.[44] The Mississippi clover, however, may well have been the plant described by R. L. Allen, of the *American Agriculturist*, as the Yellow, or Shamrock clover *(t. procumbens)*, which grew to a height of thirty inches in most of the United States.[45] The only difference between the Mississippi yellow clover and the same variety in other localities was its culture as a pasture grass. In other parts of the country, livestock grazed on it in its wild state. Only in the Natchez area in the late 1840s was it planted and cultivated.

As interest in pasture building grew among Mississippians, storekeepers in Vicksburg and Natchez added Northern grass seeds to their stocks of agricultural merchandise. The firm of S. Garvin and Company advertised blue grass, herds grass, and clover for sale as early as 1841. Timothy, orchard grass, alfalfa, mesquite grass, and numerous others subsequently were added to the list of seeds available in Southwestern stores and commission houses. Eventually Mississippi planters were able to buy in Vicksburg, Natchez, Yazoo City, or Columbus, any variety of grass seed that could be found in New Orleans, Baltimore, or New York.

Yet grasses of Northern origin did not become important in Mississippi during either the 'forties or the 'fifties. When Southwestern cotton growers planted grass for summer pasturage, they generally sodded their fields with Bermuda grass, a plant long familiar to inhabitants of the old District of Natchez.[47] No one knows when Bermuda grass was first introduced into Southwest Mississippi. Legend says that this grass was brought into South Carolina during the Revolutionary War as stuffing in the saddles of British cavalry; and it may also have been introduced from the West Indies into Mississippi by the British about the same time. Or it may have been imported from Santo Domingo by the French. In any case, Bermuda was well known in the coastal region of the Lower South.[48]

About the time the movement for agricultural improvement rose in the Eastern cotton states, South Carolinians and Georgians learned that Bermuda would grow on impoverished hill sides thickly enough to prevent soil erosion, and that—if adequately watered—it would provide excellent grazing for stock. True, the grass tended to die down in periods of drought, but it always recovered rapidly after the first rainfall. Bermuda appeared early in the spring and lasted until killed by frost in the autumn, and thus afforded grazing for some six months of the year.

In the early 'forties Martin W. Philips, N. G. North and Thomas Affleck undertook, in their articles for newspapers and magazines, to bring the advantages of Bermuda grass to public notice. Affleck was particularly effective in publicizing Bermuda pastures and hay. In 1849, for instance, he reported to the U. S. Patent Office: "I have imported from Europe seeds of over forty kinds, from Texas and the Far West over ten or a dozen, and

have also tried any number of native grasses with varied success
. . . but . . . after careful and repeated trials, I have found no
grass to compare for hay or pasture with . . . Bermuda grass." [49]
Its yield, he continued, was greater than that of any other grass
he had ever seen, and it was unexcelled in "sweetness and nutri-
tive qualities." His own experience, Affleck said, showed that
five tons of hay to the acre and even more could be cut from
Bermuda on rich soil and that as many as three cuttings could
be obtained in seasons of ample rainfall.

On the other hand, Bermuda proved to be a pest in cotton
and corn fields. Once it had taken hold, its thick, tough sod was
difficult to break with the light plows then in use in Mississippi;
and eradicating it was a formidable problem. It could not, therefore, be used as part of a cotton, corn, grass sequence in a
scheme of crop rotation. It was useful only as an erosion pre-
ventative on hill sides too steep to cultivate or on fields that had
been permanently assigned to pasturage. Many Southwestern
farmers and planters made the mistake of planting cotton or
corn on fields which had previously been sodded with Bermuda,
believing that this grass enriched the soil in much the same way
as clover or cowpeas.[50] When they ran into serious trouble in
cultivating grass infested crops, they began to write public letters
attacking Bermuda as a dangerous nuisance. As others who had
learned to value the grass as a means of combatting erosion and
a source of feed rallied to its defense, the resulting controversy
raged unabated throughout the remainder of the ante-bellum
period.

From their different points of view, both sides in the Bermuda grass argument were right. The grass was troublesome in
cotton fields, as planters contended. It also was an almost un-
rivalled pasture grass, as the livestock breeders insisted. On the
whole, the supporters of Bermuda seem to have eventually
gained a shade the better of the ink-strewn battle. Consider, for
example, an extract from a typical letter written by one of Bermuda's proponents. The author of this communication was a
small farmer of the Milldale community in the region of Vicksburg, and he customarily wrote for the agricultural press under
the pen name of "Bermuda Grass."

All I want to know is how those gentlemen who are
so bedeviled by the critter, Bermuda Grass, managed to
get it to spread so. I have been working with it now
three years and have only got it over about four acres
of ground. This I prize very highly, for it keeps from
twenty-five to thirty calves in good order all season.
These calves I would have to turn out if it were not for
this little patch of Bermuda. I would then be without
milk and butter with which my family is supplied in
profusion, besides a hundred and forty or fifty dollars
which my better half obtains by sending butter to
Vicksburg. . . .[51]

During the 'fifties a growing number of cotton growers and
stockmen were converted to the view expressed by "Bermuda
Grass." By 1860, the much maligned Bermuda had become virtually the standard source of summer pasturage on Mississippi
plantations where good livestock were valued.

The 1840s were years in which many orchards were planted
in Mississippi. Ever since the beginnings of white settlement,
several kinds of fruit had been raised in the Natchez District
and along the Gulf Coast. Nevertheless fruit trees were scarce
and orchards rare in settled sections of the state. During the
land rush era, settlers usually had been too occupied with
clearing "new ground" for cotton and corn to bother with non-
essentials. Besides many new cotton plantations then being put
into production were the property of absentee owners who had
little or no interest in the welfare of their Negroes beyond safeguarding their ability to produce as much cotton as possible.
Hence, so long as cotton brought high prices, orchards or vine-
yards were seldom planted in Central or Northern Mississippi.
After the collapse of the cotton market, however, fruit received
a share of the attention that cotton growers generally gave to
foodstuffs. Numerous orchards of apples, peaches, pears, and
plums were set out in the Indian Cession country, especially
on farms and plantations where proprietors resided with their
families.[52] In all but a few instances, the fruit from these
orchards was consumed at home. It brought little or no cash

income to Mississippians. On the other hand, fruit added variety to the monotonous diet of both Negroes and whites, improved their health, and raised their standards of living.

Orchard improvement on a considerable scale created a new market for young fruit trees in Mississippi. Southwestern planters at first were obliged to rely upon stock obtained from nurseries in the North, but they usually found Northern trees to be unsatisfactory. Young trees raised in a colder climate often died a short while after being planted in the Lower Mississippi Valley, and those that did survive were seldom heavy producers. In this situation, supply inevitably followed demand. During the early years of the depression, several professional and amateur horticulturists founded nurseries near Natchez and Vicksburg to raise acclimated fruit trees for sale to local planters.

The Southern Nurseries at Washington were the best known of the Mississippi establishments. These, as previously noted, had been founded by Thomas Affleck when he settled in the Natchez region in 1842. Skill and several years of the hardest labor were required to build up a nursery, plant the nucleus of an orchard, and restore fertility to the eroded red clay hills of "Ingleside Plantation." [53] Eventually, however, the stubborn Scottish horticulturist succeeded, and he offered for sale many varieties of young trees and fruit. Both found a ready market in the Old Southwest. Judicious advertising in Southern agricultural publications brought orders from customers in all of the cotton states. Early success encouraged Affleck to enlarge his business, particularly the nursery. By 1849, he was selling apple trees in three hundred varieties, pears in two hundred, plums in forty, and a large assortment of quinces and apricots.[54]

While Affleck was engaged in building up the Southern Nurseries, an interesting and almost unique fruit-and-cotton plantation was taking shape in Central Mississippi. This unusual establishment, which featured long rows of different kinds of fruit trees running through fields of cotton, came into being as a result of chance. Its owner, John Hebron, had migrated to Warren County from Virginia at the height of the land rush into North Mississippi. Hebron brought peach seeds from Virginia, which he and his seventy-odd slaves planted at regular intervals in cotton rows, so that they could get the benefit of cultivation given to the cotton. The heavy fruit crops these trees bore were shared by the Hebron family, their slaves and their hogs, until Hebron learned by accident that peaches were commanding good prices in New Orleans. He got this information from a free Negro who purchased ten barrels of peaches from the planter and then sold them in New Orleans. When the Negro returned to Bovina for more fruit, Hebron learned that the peaches he had sold for five dollars a barrel had brought twenty-nine dollars. This news opened his eyes to "a better business than raising cotton." [55]

In the next few years the former Virginian added so many peach, pear and plum trees to his original orchard that they ultimately covered several hundred acres. He sought the maximum number of varieties of each fruit in order to have as continuous a supply as possible for sale during the summer and fall. He also built up a huge nursery of young trees from which to supply his own wants and those of his neighbors.[56]

As several years were required to bring new trees into bearing, the Hebron orchards did not reach the peak of their production until the close of the ante-bellum period. Nevertheless, by 1854 some fifteen hundred pear trees, in addition to the original peach orchard, were yielding fruit for the market; and the sale of peaches and pears was averaging six hundred dollars an acre. At that time the orchards contained ten thousand trees, and the nursery another seventy thousand. Hebron was shipping pears largely by steamboat to markets as distant as New Orleans, St. Louis and Chicago; and he was sometimes called "the pear King." [57]

The success of this pioneer Southwestern venture into commercial fruit growing appeared assured at the time of Mississippi's withdrawal from the Union; actually, however, its sands were running out. In 1863 the Hebron orchards and residence were destroyed in Grant's final campaign against Vicksburg. Hebron died in 1862 and therefore did not witness either the ruin of his orchards or the capitulation of his beloved city.

Mississippi's other commercial orchards and nurseries of the 1840s—including Hatch and Company, at Yazoo City; the Lambart Nursery, at Vicksburg; the Starkey Nursery, at Port Gibson; and a fourth establishment at Edwards, belonging to M. W. Philips—were all much smaller than either Affleck's or Hebron's. Nevertheless, they all enjoyed a prosperity which, though modest, was substantial enough to bring a number of additional competitors into the field in the next decade. In 1860 the combined financial returns from the dozen-odd nurseries were infinitesimal when compared with income from cotton. Yet these horticultural concerns were more important than income alone indicated. They were opening up new agricultural opportunities to the farmers and planters of Mississippi; and their example might well have led to the development of a second cash crop for the state had the Civil War not put an end to their endeavors.[58]

During the closing years of the ante-bellum period, a new crop gave promise of improving the diet of the lower classes almost as much as fruit. This was the cane, sorghum, which the U. S. Patent Office imported from Asia and introduced into the South in 1854. Many Mississippians grew experimental plots in 1855 and 1856 using seeds provided by the government, and all who tried the cane were enthusiastic about its yield, its value as stock feed, and especially its molasses. The latter, which could be made inexpensively from sorghum juice, was a substitute for cane syrup and sugar. Sugarcane could not be produced in the latitudes of North Mississippi, and sugar and syrup, as a result, were too expensive for regular use by any but the wealthier people. The demand for a cheap sweeting to vary the common diet of pork, sweet potatoes, and corn bread was almost universal; consequently, the cultivation of sorghum for domestic purposes spread rapidly. By 1858, patches of the cane were to be seen throughout the cotton growing region, and sorghum molasses was on almost every table in the state.[59]

By 1849 Mississippians had largely achieved their ambition to free themselves from dependence upon the Upper Mississippi Valley for food. Neither small farmers nor large planters were any longer importing corn except after seasons of drought or excessive rain. A great majority were raising and curing enough pork to meet most of their own needs. Most livestock owners no longer had to import corn fodder and hay, because they had increased their own production of fodder and had built up grass pastures for spring, summer, and fall grazing. By experiments with Egyptian oats and rye, some were trying to develop green feeding grounds the year round. These experiments sought to close the gap left between the fall peavine field and the spring crops of oats, millet, and Bermuda grass. Many farmers and planters were also making improvements in the diets of their families and slaves by planting orchards, sorghum, and larger vegetable gardens, and by raising more wheat. Further changes in the direction of economic self-sufficiency were made, of course, between 1850 and 1860, but they were largely continuations of trends begun during the depression years.[60]

CHAPTER VIII

Mississippi Cotton Breeders

To most of Mississippi's cotton growers the depression of 1837-49 was a financial disaster. To a few, however, who were shrewd enough to recognize favorable developments and able enough to capitalize upon them, the period was one of opportunity. Among these, it will be remembered, were Colin Tarpley, who had discovered in 1840 that falling cotton prices had created a market in his vicinity for thoroughbred livestock, and Thomas Affleck, who had established the Southern Nurseries in 1842 to fill a strong local demand for trees and fruit. In much the same way, Martin W. Philips, Richard Abbey, Henry W. Vick, John Hebron, William B. Farmer, G. D. Mitchell, William Hogan, and several other Central Mississippi planters, learned after 1845 that they could augment their incomes by breeding improved strains of cotton and selling seed to customers located in all the states of the Lower South from Texas to South Carolina.

From 1837 to 1849, the average cotton planter tried to offset unfavorable cotton prices by increasing the size of his crop and improving the quality of his staple. Hoping to increase his yields, he tried to enrich his soil by rotating crops and by using such fertilizers as cotton seed, cowpeas, and animal manures. To make possible the cultivation of more land, many a planter supplied his workers with more and better horse-drawn farm equipment. In addition, farmers engaged in a constant search for breeds of cotton having longer staples and higher yields than the familiar Mexican or Petit Gulf.[1] The depression, therefore, indirectly put a premium upon the seed of superior varieties of cotton.

Three classes of agriculturists attempted to turn the mania for new strains of cotton to their advantage. In the first class were would-be seed dealers who imported foreign cottons from Asia, Africa, and Central and South America and tried to sell them—unsuccessfully on the whole—to farmers and planters of the cotton states. A second group, far more numerous, consisted of the many ordinary cotton growers who cherished hopes of balancing their budgets by selling their own cotton seed. Optimists of this category made a practice of searching their cotton fields for accidental mutations or crosses with unusual characteristics, and, when the hunt was successful, put their discoveries on the market under such supposedly descriptive names as "Chinese Silk," "Twin Cotton," or "Okra." Buyers who succumbed to the florid advertising that often marked the entrance of these name-brands into the trade, frequently were disappointed. In most cases, the promoters of the "new" cotton had selected original plants because of their unusual appearance rather than for the quality or quantity of their lint. Hence new varieties often yielded less than their proprietors claimed, and their fiber was not always as good as that of the common Mexican. Buyers who considered themselves cheated by such dealers often wrote letters to the editors of newspapers and agricultural periodicals contradicting the extravagant claims made by the originators of these expensive "humbugs."

In the third class of seed suppliers were professional plant breeders. It was they who, by deliberate hybridization and painstaking selective breeding, developed most of the really valuable strains of cotton that appeared between 1837 and 1860. The number of useful new strains of cotton thus produced was small—even though they were sold under countless different labels—for only a few Southerners were engaged in this exacting occupation. Of this little group of plant breeders, most lived in Mississippi where there was a long established tradition of quasi-scientific cotton breeding inherited from William Dunbar, Dr. Rush Nutt and his neighbors of the Gulf Hills.[2]

During the 1830s Petit Gulf cotton growers had built up a profitable trade in choice Mexican cotton seed, and they expanded their business during the depression. In the 1840s planters in the newly-settled Yazoo-Mississippi Delta and in the Central Mississippi counties of Warren, Hinds, and Madison, bought some of this seed from the Nutt group, learned their breeding technique, and joined them in the business of selling seed. These newcomers to the trade marketed their product as Petit Gulf seed, even though they lived many miles from the old Gulf Hills. Soon, therefore, the name Petit Gulf lost all geographic significance and came to represent nothing more than carefully selected seed of Mississippi Mexican cotton. Yet Petit Gulf lost none of its reputation. So long as this cotton was planted in fertile soil and its seed chosen with care, it retained all the characteristics that first won for it favorable attention.[3]

Indeed, the South-wide demand for Mississippi Petit Gulf remained strong, even after two new mutations derived from the old Mexican stock—Okra and Chinese Silk—came on to the market in the early 1840s. Though heralded by vigorous advertising, these two new strains could not supplant Petit Gulf.[4] Not till 1845 did this variety encounter serious competition—and that competition was strongest from breeds of cotton developed in Mississippi.

When Richard Abbey of Yazoo County made a tempestuous entrance into the business of selling cotton seed in 1845, he dealt it the first of a series of shocks that drastically altered its placid and noncompetitive nature. Abbey, like Thomas Hall, was a typical self-made man who found full scope for his talents in the Old Southwest. Unlike Hall, however, he came to the planter's mansion by way of the tradesman's entrance, not by the overseer's cabin. As a New Yorker turned Mississippian and a merchant turned planter, Abbey was as different from the usual seed dealer as a railroad train from an ox wagon. He was by inclination a salesman and promoter rather than a farmer, and he occasionally shocked Mississippians by his Yankee drive. Yet he taught his fellow seedsmen important lessons about the value of advertising.[5]

When Abbey arrived at Natchez as a nineteen-year old lad in 1826, he had little formal education and owned almost nothing beyond the clothes on his back. Yet he had bountiful energy, self-confidence, ambition and intelligence. He found employment with the mercantile house of Merrick and Company and quickly rose to the position of junior partner and manager. Upon Merrick's death, Abbey gained full control of the business, and for the next seven years directed it profitably. Having accumulated a modest fortune by 1840, Abbey sold the business and invested the proceeds in a lowland cotton plantation in Yazoo County.

Although Abbey knew almost nothing about the practical side of cotton farming when he bought "Boston Plantation," he did not long remain ignorant. He studied the farming methods of his neighbors closely, and read everything obtainable about agriculture. Within two years he had learned enough about cotton plantations and their problems to begin writing articles for the *Southwestern Farmer* and the *Southern Planter*. Abbey did his early writing under various pseudonyms, but soon acquired enough confidence in the soundness of his views to sign his articles "R. Abbey, of Boston Plantation." Letters he published established his position as an apt and discerning analyst of Southern agriculture. One article explaining his method of raising and preparing cotton for market, published in *De Bow's Review* in 1846, was undoubtedly the clearest description of its kind printed in that decade. It also demonstrated beyond a shadow of doubt that Abbey had become thoroughly versed in the fundamentals of cotton planting.[6]

As a result of his omnivorous reading about agriculture, Abbey developed an interest in the commercial possibilities of plant breeding. From agricultural journals he learned that some promoters of name-brand cottons had made handsome profits by selling their seed; and he concluded that what they had done, he could do better. All that the seed business appeared to require was a talent for salesmanship and a good new variety of cotton. In 1845, Abbey possessed both. Over the preceding four years, he had been cultivating and improving a cotton of Mexican origin quite unlike the common Petit Gulf strain. This plant had a large stalk and bolls fully twice as large as those of Petit Gulf. Its lint was extraordinarily fine in quality, and the fiber was almost as long as the best Sea Island cotton. The large white seeds were thickly coated with woolly linters. The outward appearance of the seed was important to Abbey. He improved his cotton by selecting planting seed with great care from the gin pile on the basis of appearance alone. He therefore gave his cotton a name suggestive of its seed: "Mastodon"—the largest and wooliest creature he knew of.[7]

Abbey saved all the seeds from his first three crops of Mastodon and used them to increase the acreage he devoted to that variety. By 1844 the crop was large enough to enable him to send packets of the seed to twenty or thirty friends to be used for experimental purposes. In 1845 the Mastodon crop on Boston Plantation yielded thirty bales of exceptionally fine lint; and Abbey decided that fall to test its market value. So he shipped the thirty bales to Buckner and Stanton, his agents in New Orleans. Although most cotton prices in 1845 were at the lowest point of the ante-bellum period—three to five cents a pound—Abbey received $1,856.29 for his thirty bales, an average of sixteen cents a pound. Some of his friends also placed small lots of Mastodon cotton on sale. Their parcels did not bring prices so high as Abbey's, but none went for less than twelve and a half cents a pound, an astonishingly high price for that year.[8]

With this incontrovertible evidence of Mastodon's superiority in hand, Abbey was ready at last to begin selling his seed. He conducted an elaborate advertising campaign in the winter of 1845-46, in order to bring Mastodon to public notice before the 1846 planting season. He wrote letters to influential newspapers and agricultural journals relating the story of his cotton breeding experiments and their ultimate success.[9] He asked for testimonials from the friends who had planted samples of Mastodon,[10] and several responded with articles for publication in their local papers.[11] He sent specimens of the stalk, seed and fiber to many New Orleans business houses for exhibition. To the Louisiana State Fair at Baton Rouge in January, he sent specimens of the plant and samples of its seed and lint, along with an open letter giving full particulars about its history and characteristics and the best methods of cultivating and preparing it for market.[12]

Abbey's 1845-46 advertising program was phenomenally successful. The extraordinary length of Mastodon (two inches in comparison to the one and three-quarters of the famous old black seed Creole cotton) and the high prices paid for it captured the imagination of Southern farmers and planters. Letters about Mastodon's advantages and disadvantages poured into the offices of newspapers all over the South.[13] Some came from planters who wanted to learn more about Mastodon cotton, and some from people who wished to know where they could buy seed. Others were from persons who feared that Mastodon's supposedly high productivity would eventually flood the market with cotton and thus depress prices even lower than they were then. Finally, reports came from a few kill-joys, like Martin W. Philips, who had tested the cotton and found that its properties were not quite so ideal as the public had been led to believe.[14] On the whole, however, Mastodon's critics were in the minority the first few years it was on the market, and Abbey and other growers sold large quantities of the seed at good profits.

After a few seasons Mastodon began to lose some of its quickly-won popularity. Planters, who had been over-sold on its virtues in the beginning were disillusioned to discover that it had disadvantages as well as advantages.[15] The staple of pure Mastodon was really as good as Abbey had claimed, and it continued to sell for slightly higher prices than Petit Gulf. Yet it had a serious liability; like Sea Island cotton, it required special care in cleaning or it was likely to be damaged by the saws of the gin. If ginned and baled with the same equipment and in the same manner as short staple cottons, Mastodon suffered a reduction in grade because of mangling of the lint and lost value proportionately. Abbey himself overcame this difficulty by making modifications in his gins, and he passed his plans on to the Carver Manufacturing Company of Massachusetts. Tests satisfied Carver that the changes were effective, and he began to manufacture the special gins in quantity. Ordinary cotton planters, not willing to go to the expense of buying gins especially for a single cotton variety, ginned their Mastodon on common machines. Consequently, they were disappointed when it sold for little more than Mexican cotton.

Cotton producers were troubled also by degeneration of the Mastodon plant after several years of steady cultivation. This too was mostly the fault of the farmer rather than the seed breeder. Very few persons in the South were aware that cotton was subject to natural crossbreeding by exchanging pollens, and even fewer were careful to preserve the purity of their strains by separating them adequately from one another. Planters who used Mastodon unwittingly allowed it to become mixed with common breeds, and it gradually lost its distinguishing characteristics. Abbey had warned his customers of this danger and had advanced methods of avoiding it. Yet many of them failed to heed his advice. When their Mastodon cotton degenerated, as Abbey had assured them that it would unless properly safeguarded, they sometimes unjustly branded his cotton as a "humbug." [16]

Despite its partial fall from grace, Mastodon continued to enjoy a limited popularity throughout the remainder of the ante-bellum period. For it had one characteristic that endeared it to the planter who sought to produce maximum cotton crops each year. Mastodon lint clung so tightly to its huge burr that winds and heavy rains seldom tore it from the pod. Thus Mastodon crops could be left standing in the fields until all other cotton had been picked. Some farmers took advantage of its ability to withstand weather by planting small plots of it which they left standing as late as January. Whatever these fields yielded was so much net addition to the farmer's total production.[17] In brief, Abbey's long staple cotton by 1860 had become a supplement, rather than a rival, to short staple cottons of Mexican descent.

Serious competition to Petit Gulf was left to Hogan's "Banana," Farmer's "Sugar Loaf" and the "Hundred Seed" cotton of Colonel Henry W. Vick, all of them refinements of Petit Gulf stock.[18]

Nothing in the career of Henry W. Vick, of Warren, Madison and Yazoo Counties, before 1839 suggested that he would eventually become the most successful cotton breeder and seed salesman of the Old South. He had inherited a moderate fortune from his father, Reverend Newet Vick (for whom the town of Vicksburg was named), and had added to his patrimony during the boom years of the 1830s. In those years and afterward Vick was a typical member of the large slaveowning class. He owned and operated several plantations in the hills of Central Mississippi and in the Yazoo-Mississippi Delta country. Their management he customarily left to overseers, limiting his own interference to periodic visits of inspection. Like most absentee landowners of the time, Vick was chronically dissatisfied with the work of his overseers. Yet his discontent was not enough to cause him to take personal control of his plantations. In common with many planters, he was more concerned with "internal improvements"—especially Mississippi River levees—and the fortunes of the Whig party than he was with agricultural reform. That such a man should have revolutionized Mississippi cotton breeding methods and succeeded in developing one of the Old South's most famous brands of cotton, was itself extraordinary.

Vick's original stock of cotton seed had been "little brown" (or drab) Petit Gulf Mexican, which he obtained from the Gulf Hills near Rodney. For seven years he cultivated this strain without making any attempt to improve or modify its characteristics. He did take care to preserve its purity by selecting uniform seed for planting, but that was the full extent of his activity in cotton breeding between 1830 and 1837. For some reason, he ultimately became dissatisfied with his Petit Gulf cotton, and, in the year of the panic, set about improving it by the method of selective breeding introduced into Mississippi by Rush Nutt. Following Nutt's system, Vick instructed his overseers to choose planting seeds having a certain characteristic appearance. The theory upon which the Nutt system was based, of course, was that uniform seed would produce uniform cotton; its corollary was that superior seed would produce superior cotton. There was an inherent difficulty, however: how was one to tell which were the superior seeds, if they were judged only by appearance? After two years of experiments, Vick concluded that superior seed could not be detected by size, shape, or color, and he discarded the Nutt approach altogether.[19]

After considerable thought, Vick decided to try a method of selective breeding which had long been used effectively by Northern corn growers, but which at the time was almost unknown to cotton planters of the South. It consisted of selecting the healthiest and most productive plants in the field rather than the prettiest seed in the bin. In applying this method to cotton growing, Vick instructed his overseers to send his most intelligent Negroes into the field in advance of the regular picking gangs. These Negroes were to gather seed cotton from nothing but the largest and most productive cotton plants. This cotton then was ginned separately from the rest of the crop, and its seeds reserved for the following year's planting. Vick was temporarily satisfied with the improvement obtained in this way, continuing the practice without modification from 1839 to 1844.[20]

In those years Vick generally placed his stock of surplus seed on consignment with merchants in Vicksburg, Yazoo City, Jackson and Natchez. In the early 'forties he was pleased to discover that his seed were rapidly acquiring a reputation for high quality. In fact, many cotton growers were willing to pay slightly higher prices for Vick's seeds when they learned that his pickers were gathering them "from the best bolls from the best stalks only." [21]

In 1843, Vick became sufficiently interested in the plant breeding experiments being conducted on his plantations to take a more active hand in the enterprise. That year he happened to spend the summer and fall at "Nita Yuma," a plantation located on Deer Creek in Yazoo County; and he whiled away the dull Delta evenings by examining cotton specimens which he had gathered during the day from plants of unusual size and yields of lint. When he began to inspect these locks with minute care, Vick made a remarkable discovery that became the starting point for his later experiments. This was the story, as he related it:

> In examining the products of the different stalks, which amounted to hundreds of bundles, and was the labor of a winter, I spread the bundles out on the table before me. Satisfied that the fingers would bring to light much valuable, perhaps essential, information that would be hidden by the gin, I determined to make the investigation thorough and complete. It was not long

before I plainly saw that what I had supposed to be a homogenous [sic] stock of cotton seed, consisted in fact of ten or a dozen distinct varieties. I became attentive to the shades of difference. Of these ten or a dozen, six were evidently greatly superior to the rest; my selections were confined to those. Of the six, one consisted of twelve locks only. The size and beauty of the locks, the style in which they were put up, the abundance, length, fineness and lustre of the lint, form and hue of the seed, led me to pronounce them at once a new, distinct, and valuable variety. I picked them. They yielded precisely 100 seed. Having so many names to furnish, the singularity of the incident suggested a name that would sufficiently distinguish them, and at the same time, with me, perpetuate their history. I called them my "100 Seed" cotton.[22]

In this way perhaps the best known name-brand cotton of the late ante-bellum South acquired its separate identity.

In 1844 Vick planted his tiny supply of Hundred Seed and cultivated it with exquisite care. After the cotton had ripened, he gathered and cleaned it by hand. To his delight, the lint was uniform and exactly like the splendid specimens of the year before. The seed from the 1844 crop served to plant two and a half acres in the spring of 1845. That fall, the small plot of Hundred Seed yielded two bales of splendid cotton, which sold in New Orleans for 2⅛ cents a pound more than his crop of select Petit Gulf.[23]

Vick personally went through his 1845 crop of Hundred Seed as soon as it matured gathering superior specimens for further experimentation. These he subjected to the same type of close examination that had produced the parent stock. "Judge of my surprise," Vick wrote later, "at finding my '100 Seed' springing upon me four new varieties the first year, each possessing some peculiar excellence of its own." To these fresh strains, he gave the names "Sub-Nigri," "Belle-Creole," "Diamond," and "Lintonia."[24]

The 1845 crop of Hundred Seed produced enough to plant two-thirds of Vick's cotton acreage the next spring. So that he might compare his own cotton with other name-brands of the time, he planted the remainder of his land in Abbey's Mastodon, Weem's Guatemala, Lane's Yucatan, Farmer's Sugar Loaf, Lewis's Prolific, and the Mexican-Egyptian hybrid developed by Haller Nutt. When the crop had been gathered, ginned, and examined, Vick was fully convinced of the superiority of Hundred Seed.[25] He then began to sell seed of his improved variety at $1.00 to $1.50 a bushel. The price depended upon the quantity ordered, but even $1.50 was more reasonable than the charges usually made for new varieties.[26] The low cost and uniformly high quality of the Vick Hundred Seed gradually made it a favorite with many Southern cotton growers. For the remainder of the ante-bellum era, customers from all the cotton states continued to buy everything Vick could supply.[27]

In this era of unbridled capitalism, no legal or ethical barriers prevented a seed salesman from appropriating the work of another as his own. Consequently, cotton growers sometimes obtained seed of improved varieties from a breeder, grew crops of their own from them, and then sold the seed under the original owner's brand or under other names of their own choosing. Such practices often led to abuses costly both to the customer and the developer of the cotton. If the suppliers retained the cotton's true name, the reputation of the originator was likely to suffer. For seed sold by such persons were frequently inferior to those supplied by the original owner. The customer, of course, was liable to financial losses when he bought poor seed at premium prices. In cases where the seller gave a name of his own selection to the cotton, the seed breeder's reputation was protected, but the customer was not so fortunate. Thus a planter ran a risk when he purchased name-brand seed unless he was personally acquainted with the supplier. In time this unfortunate situation was resolved by the development of "pedigreed" seed, of which the salesman guaranteed the purity and parentage. But this solution to the problem was not found in time to be of much help to seed breeders or cotton planters of the pre-war period.

Vick, as the leading cotton breeder of Mississippi, was the one most frequently victimized by seed dealers. "Jethro" cotton illustrated how other men profited from his work. In the course of experimenting with his 1846 crop of Hundred Seed, Vick sent several different samples to his old friend Martin Philips for testing in the field under new conditions. Philips planted them in an experimental plot in the spring of 1847 and cultivated and picked the cotton with his own hand. He had different cottons ginned separately and then compared them with many other kinds that he had in his possession. After tabulating the results, Philips sent a lengthy report to Vick generally very favorable to his strains. For further investigation, Philips kept a portion of the seed from Vick's cottons. The remainder he distributed among such friends as ex-Governor James Henry Hammond, of South Carolina, and Jethro V. Jones, a former editor of the *Southern Cultivator*. Philips labelled the package of Vick seed which he sent to Jones with the name "Jethro" cotton as a compliment to a fellow worker in the field of agricultural improvement. Jones planted the seed on his experimental farm in Burke County, Georgia, found the Vick cotton exceptionally well suited to conditions in Middle Georgia, and subsequently, through the medium of the *Southern Cultivator,* made "Jethro" famous among planters of the Southeastern states. Vick, the originator of the strain, received no financial benefits whatever from those sales in the Southeastern states. Philips, however, in several letters to the press carefully publicized the facts about Jethro's history, and he gave full honors to Vick for his pathbreaking work.[28]

In the development of at least four other cotton strains, Vick apparently did not receive the credit due him. In all likelihood, Banana, Cluster, Hogan, and Pomegranate, (four well-known brands of the 1850s) were offsprings of Vick's Hundred Seed. These cottons were placed suddenly on the market by several of Vick's Warren County neighbors, none of whom had any previous reputation as a cotton breeder. The fact that these strains proved to be identical and of high excellence indicates a common origin in the vicinity of Warren County. Significantly, not one of the promoters ever discussed the ancestry of his strain.

"Banana Cotton" was introduced to the public by William

Hogan and John Hebron of the Bovina community of Warren County.[29] It won almost immediate acceptance from planters in all the cotton states, and by the end of the decade was being sold under as many labels "as there are persons who desire to make money by selling seed." [30] Banana is known to have gone out from Mississippi to other states as Boyd's Cluster, Boyd's Prolific, Washburn's Olive, Hogan, Hebron's Banana, and Mitchell's Pomegranate. There doubtless were many other aliases that have long since been forgotten.[31] The price of the seed varied far more widely than the price of the lint, and depended upon the greed and salesmanship of the supplier. Martin W. Philips sold Banana seed regularly at one dollar a bushel;[32] Mitchell sold Pomegranate in Georgia and South Carolina in 1850 at $2.50 a bushel; and John Hebron and William Hogan at one time marketed Banana at ten cents a seed.[33] On the whole, however, Banana could be obtained for $1.50 a bushel during most of the 1850s.[34]

No matter what its name, Banana was a fine and productive variety of Mississippi cotton, and well deserved its extraordinary reputation.[35] In 1850, for example, planters of Sumter County, Alabama, conducted tests with Pomegranate seed obtained from General George D. Mitchell, of Warren County, Mississippi. Sumter County soil was much less fertile than either the hills of Warren County, Mississippi, or the lowlands of Yazoo County, and the stands of cotton obtained by the Alabama farmers were below average. Yet on their test plots they produced averages as high as two thousand pounds of seed cotton to the acre.[36] The stalks of Pomegranate cotton grew to heights of three feet on poor land and six to eight feet on rich bottom lands, and were thickly covered with big bolls. Philips, on one occasion, counted as many as fifty-four bolls on a bush two and a half feet tall, and 134 on a three foot stalk.[37] Pomegranate's staple was somewhat longer than that of ordinary Petit Gulf and consequently brought a better than average price at New Orleans.

The business of selling name-brand seed proved highly profitable to such planters as Martin W. Philips, who knew how to advertise effectively. Philips's agricultural writings bore a high reputation throughout the Southern states, and agricul-

turists were accustomed to turning to him for advice. As soon as
new varieties of cotton began to appear on the market in the
'forties, farmers directed inquiries to him about seed, and some
asked him for samples. In this indirect manner Philips was
drawn into the business of selling seed. He conducted tests with
all the different seeds he could obtain, in order to answer questions from his agricultural friends; and eventually began to sell
surplus seeds. Because Philips frequently received orders that
he could not fill from his own supplies, he became a middle man
for other cotton growers from whom he procured seed. This seed
trade grew rapidly in the 1840s and finally became a major
source of income. By 1850, Philips had made enough profit
from the sale of seed to liquidate debts incurred at the time of
the crash.[38]

While Philips developed no new varieties of cotton, he performed useful services to cotton growers by testing dozens of
new strains of cotton on his Hinds County plantation. He issued
truthful and unbiased reports on the comparative worth of these
strains, and waged a long and sometimes acrimonious fight to
bring honesty into the Southern cotton seed industry. The
reforming journalist always insisted that dispensers of seed use
correct and accurate labels and tell their customers where the
seed came from. He recommended that each distinct variety of
cotton be given a standard name that would be the same regardless of the location or ownership of the plantation where it was
grown. In the 'fifties, Philips fought more than one battle to
protect the pocketbook of the customer. In one well publicized
episode, General George D. Mitchell, a planter of Warren
County, had the misfortune to become the target of Philips's
wrath. In 1850 Mitchell began a most effective advertising campaign to convince planters everywhere of the superiority of his
brand of cotton.[39] He succeeded so well that he soon sold out
his supply of Pomegranate seed. Because his customers were
paying $2.50 a bushel for it, Mitchell was loath to return any
unfilled orders. So he purchased Banana seed from neighbors
for a dollar a bushel and resold it as Pomegranate for more than
twice as much. There was no deliberate fraud, for Mitchell,
Hebron, Hogan, Philips, and other seed producers were aware

that Banana and Pomegranate were one and the same. Philips, however, was incensed by Mitchell's high prices, and he proceeded to expose the General's profitable scheme. Curiously, Mitchell's customers still persisted in ordering Pomegranate seed from him at his price even after Philips had declared publicly that they could buy it elsewhere for as little as a dollar a bushel. This certainly was a demonstration of the power of advertising— or of the persistence of suckers![40]

Although Philips failed to spike Mitchell's guns completely in this particular instance, his reports probably did have the effect of checking others who might have been tempted to exceed the latitude enjoyed by seed breeders of the time. And his commentaries on the relative value of different varieties possessed value for honest breeders as well as for their customers.

In addition to other services to seed breeders, Philips tried to popularize information about hybridization. Of course, he did not originate these principles, which had long been known to botanists. Nor was he the first cotton planter to conduct experiments in crossing cotton breeds. Yet Philips did state those principles in terms that Mississippians could understand in an article published in the Yazoo City *Yazoo Democrat* in 1851. In part Philips said:

> I think cotton will mix and that thus a cross breed may be raised—but only if done through the bloom. My reason for stating this plainly is that I have been asked if there could be mixture by roots. I think it is possible only that wind might blow the reproductive matter from the plant upon the proper organs in another plant—and thus fructification might take place. But this is barely possible from the shape of the bloom. I feel well assured that this does take place by the means of bugs, flies, or insects in general, passing from one bloom to another—but not to the extent that many suppose . . . so soon as the bloom opens . . . the stamens (male apparatus of a flower) burst and give out the dust, which is received by the pistil (the female organ) and is thus fructified. The pistil, or 'pistillum,' contains the

'ovules,' or young seed within the ovary and must be fructified by the dust contained in the stamens. Thus a bee if entering the blooms early enough, may bring about a hybrid. . . .[41]

Unfortunately, Mississippi cotton breeders made little use of the technique of crossing plants which Philips suggested. They preferred instead to depend either upon the old methods of Rush Nutt or upon the better system of plant improvement worked out by Henry W. Vick.[42] On the whole, Mississippians took little or no advantage of botanical knowledge until after the Civil War.

By the close of the ante-bellum period the Southern cotton seed trade had undergone radical changes. Where improved Mexican seed had sold anonymously as Petit Gulf during the late 'thirties, they were marketed in the 'fifties under brand names peculiar to the supplier. Competition in price and advertising had become the rule, and customers demanded that their planting seed be selected by the breeder in accordance with the system developed by Henry W. Vick. In brief, the trend was toward longer staples and higher per acre yields than Petit Gulf once had provided. A class of professional seed breeders and suppliers had come into being to supply the demand for cottons of specialized types. Only the wide use of hybridization techniques and the certification of seed were needed to transform the cotton industry of the late ante-bellum period into its modern counterpart.

CHAPTER IX

Cotton Production During the Depression

Mississippians were puzzled by the downward trend of cotton prices during the 1837-45 period. Convinced that the world-wide market for textiles was still growing, they at first were unable to comprehend that their staple crop was in excess of supply. Instead, they blamed their economic difficulties upon commission merchants, shippers, and textile manufacturers, whom they suspected of conspiring together to control the cotton market. The conspiracy theory was widely held for a time in the Lower Mississippi Valley, and it led a small number of Mississippi bankers, merchants and planters into an unfortunate venture into the export trade. They organized several shipping companies and erected steam compresses and warehouses in the river ports of Natchez, Grand Gulf and Vicksburg. Ocean-going sailing vessels were towed up to these points by tugs, loaded with re-compressed cotton and drifted downstream to the Gulf. There they spread their sails for overseas destinations.[1]

This attempt to by-pass the monopolists at New Orleans ended in financial failure. Thousands of bales were shipped to Europe safely between 1837 and 1840. Once in Liverpool or Marseilles, however, Mississippi cotton brought little better prices than in New Orleans. Furthermore, the savings effected in shipping and handling charges were not great enough to compensate for the expense of towing the sailing vessels up river. Thus direct trade proved to be unprofitable and disappointing. In the 'forties Mississippi cotton began to flow once more through the normal outlets at New Orleans and Mobile. Southwestern cotton planters subsequently were compelled to seek their economic salvation elsewhere.[2]

Although Mississippi Valley cotton growers never lost their distrust for New Orleans cotton buyers and commission merchants, some of them eventually adopted the view that overproduction was at the root of their troubles. There was disagreement, however, over the best treatment for the disease. Some wanted to raise prices by limiting the exportation of cotton. Others advocated the directly opposite solution, arguing that falling prices should be offset by producing larger crops.

A majority of Mississippi's editors and many of her larger planters subscribed to the first point of view. One editorial in the Panola *Lynx* published in 1845 was typical of a vast body of newspaper opinion. "We trust," pleaded the editor, "that none of our planting friends will be induced by a slight improvement in the price of cotton . . . to change, for a moment, their determination to diminish the amount of the production."[3] Big planters of a similar mind had a spokesman in Richard Abbey. In letters to agricultural periodicals, Abbey urged Southerners to reduce their crops and improve the grade of their cotton. The time, land, and labor conserved in this manner, he argued, could be devoted profitably to raising "a good crop of corn, potatoes, peas, etc., and for *feeding it out* to a plenteous drove of cattle, horses and hogs; nor need the spinning and weaving and various other matters of plantation economy be neglected."[4] In 1845 he spoke before a planters' convention at Jackson in favor of a plan to curtail the production of cotton through cooperative action.[5] Thomas Affleck, writing for the Albany (New York) *Cultivator* in 1844, expressed an opinion similar to Abbey's:

> That every effort should be made by the South to introduce other staples than cotton is very certain. The over-production is so great that prices cannot improve; at present prices this troublesome crop cannot be grown with profit. . . . And the only possible means of lessening this over-production is to induce the cotton

planter to turn his attention to . . . other crops . . .
even if they pay no better than cotton does now.[6]

Most of Mississippi's cotton growers, however, rejected the
course of action suggested by Affleck and Abbey. In fact, a
report to this effect was sent to the *American Agriculturist* by
Martin W. Philips in 1846. "I know it is impossible," he said,
"to persuade the planters of the cotton region especially . . .
that they have any interest equal to the [production of] the
present full [sized] crops." They should improve their economic
condition, he said, by reducing costs of operation and conserving
the fertility of the soil.[7]

John C. Jenkins, Jr., the editor and proprietor of the Vicksburg *Sentinel*, was one of the few newspapermen of Mississippi
who shared the opinion of the majority. Fearing loss of markets
to foreign competitors more than falling cotton prices, Jenkins
argued in favor of greater efficiency in cultivating the staple,
and opposed all schemes for reducing cotton acreages. To follow
the plan of Abbey and Affleck, he wrote on numerous occasions,
would be to play into the hands of English monopolists who
were trying to develop alternate sources of supply for raw cotton
in tropical colonies of the British empire. It would be better for
the Southerner, Jenkins thought, to increase rather than diminish
his output of fiber. The South must retain its share of the international trade even at the cost of accepting yet lower prices for
its chief product. To do otherwise would foster the growth of
disastrous competition from cotton producers in India and
Egypt.[8]

By their actions Mississippi agriculturists demonstrated that
they agreed with Jenkins instead of Affleck and Abbey. Most of
them attempted to raise more and more cotton. At the same
time, they tried to cut production costs by growing their own
foodstuffs and improving the productivity of their labor forces.
Furthermore, during the 'forties the settlement of the northern
half of Mississippi greatly expanded the state's cotton producing
area. It was in this decade that the hills of North Mississippi
were settled and brought under the plow, and hundreds of

plantations were cleared along the water courses of the Yazoo-Mississippi Delta. Moreover, the completion of a railroad between Vicksburg and Brandon in 1839 opened up a wide belt of land in the central part of the state. This railroad provided Warren, Yazoo, Hinds, Madison and Rankin counties with an outlet for cotton much more convenient than the Yazoo, Big Black and Pearl rivers. In the area served by the railroad, cotton production grew rapidly. In 1847, for example, the line hauled 39,901 bales; six years later it transported 97,868 bales.[9] Because of the increased acreages devoted to cotton and because of expanded production on individual farms and plantations, Mississippi cotton fields grew noticeably during the depression years. In 1836 the state exported 317,783 bales.[10] Three years later, the crop was 483,504 bales.[11] In 1849 it totaled 484,292 bales despite unfavorable weather which reduced the harvest by at least a third.[12] In brief, Mississippians reacted to falling cotton prices by growing bigger crops.

Nearly all cotton growers used the same method of increasing their production of the staple. For the first time in the history of the Old Southwest, planters began to farm more intensively. Most of them planted improved strains of cotton with longer staples than Petit Gulf, and adopted many labor-saving devices. Furthermore, Mississippians began to fertilize with legumes and cotton seed on a large scale. Barnyard manures and commercial fertilizers, however, were seldom used, or not used at all, as they required too much labor or cash.[13]

Although agricultural reformers unanimously agreed that the overseer system of management for plantations was grossly defective, very few planters of Mississippi did anything to rectify the evils that they loved to complain about. Even the reformers themselves were remiss. Philips, for example, was a constant critic of the system—although a defender of individual overseers. Yet he continued to use white employees to supervise his slaves throughout the later 'forties and the 'fifties. In practice, Philips tried to remedy an admittedly bad situation by hiring the most competent men available. The average planter did no more than the "sage of Log Hall" to improve the management of his plantation. In fact, none originated any development worthy of note in the field of labor and management during the depression or in the ensuing decade of prosperity.[14]

Among the agricultural achievements of the 1840s, none was more valuable than the improvement made in farm implements. Sod-breaking and turning plows became more durable and efficient, and several valuable new implements were developed or adapted—invented would be too strong a word—to perform specific tasks in the cultivation of row crops like cotton and corn.

The wrought iron plows commonly employed in the Lower Mississippi Valley during the 1820s and 1830s had been crudely planned and constructed, and consequently were quite inefficient in operation. Most of them were hand-made by Kentucky or Ohio blacksmiths, though in all the cotton growing states some were produced. Their wrought iron shares could be sharpened easily by plantation smiths, but quickly became dull again and soon wore out. Indeed, plows of this type seldom lasted more than one season. More important, wrought iron plows, because of poor design, were usually incapable of opening furrows deeper than two or three inches.

In the early 'forties, wrought iron plows began to give way to vastly superior implements made of cast iron.[15] Cast iron plow points and moldboards, which were made by pouring molten metal into molds, were harder to fabricate than old-fashioned wrought iron parts. The process of mold-making was too complicated for most blacksmiths, so local Southern plowmakers were forced to purchase their metal parts from foundries. As a result of employing factory-made parts, plows became better in quality. True, cast iron points would break where wrought iron ones would merely bend. Yet they retained their edges better and lasted much longer. Being better designed, the new cast iron plows performed much more satisfactorily than their predecessors. When sharpened and in good condition, cast iron sod-breaking or turning plows cut furrows four inches deep, and still were easier for teams to pull than wrought iron implements opening no more than a two or three inch cut.[16]

Martin W. Philips and Thomas Affleck were principally responsible for introducing Northern factory-made plows to

Mississippians. In the 1830s Philips had become interested in reports of improved implements published in agricultural journals, and in 1838 began systematically to import and test new plows. His early experiments revealed that Northern farm implements had defects as well as advantages where Southern needs were concerned. Because Northern sod-breaking and turning plows were made for thicker turfs than those of the South, Philips found them to be heavier than cotton planters liked. Their beams were placed too low on the stock to fit Southern harnesses properly, an arrangement which tended to raise the plows out of the ground. On the other hand, Philips found many valuable characteristics in Northern plows. They usually were made of good materials, and their design and workmanship were superior to those made in Mississippi blacksmith shops. Because of these factors, the performance of Northern factory-made plows was vastly superior to Mississippi-made implements. They opened deeper and wider furrows than Southern cast iron plows, and their draft was smaller—though their weight was greater—than the best products of Kentucky and Ohio blacksmith shops. Admittedly they were more expensive than plows made in the South—and this was often the decisive factor with depression-ridden Southern agriculturists. Yet in the long run they actually cost less. Philips proved conclusively that Northern plows lasted at least three times as long as plows turned out by local smiths and required very little maintenance. One new twenty-five cent point each year was usually all that a good Northern plow needed.[17]

Philips, who was a mechanic of some ability, made and tested a number of alterations in Northern plows.[18] He discovered that these implements performed better in light Southern soils when their beams were raised, and their weight reduced by substituting lighter parts where practicable. As tests proved that their points and moldboards were quite satisfactory, he made no attempt to modify them. Philips communicated the results of his tests and modifications to several Northern factories, and several of them subsequently incorporated his suggestions in plows manufactured for the Southern trade.

As long as he was conducting tests, Philips considered it his missionary duty to inform his neighbors about superior farm equipment. He took pleasure in demonstrating implements to visitors at Log Hall, and sometimes held public exhibitions of new plows and farm machinery. Through articles for newspapers and agricultural journals Philips made his test data public, and his views on farm machinery were received with respect in many parts of the cotton states. Knowing his influence with readers of the agricultural press, plow manufacturers both North and South were anxious to procure his endorsement. Implement makers frequently shipped new models to him without charge, hoping that he would recommend them to his readers. Thus Philips became familiar with new farm equipment almost as soon as it appeared upon the market, and his reports on their performance made him the foremost Southern authority upon the subject.

Thomas Affleck's efforts to promote the adoption of improved Northern farm implements and machinery were largely confined to the Lower Mississippi Valley. Having become familiar with the characteristics and performance of factory-made plows while on the staff of the Cincinnati *Western Farmer and Gardener,* Affleck felt less need for experimentation with farm equipment than Philips, and consequently he conducted fewer comparative tests. Instead, the journalist concentrated his efforts upon familiarizing planters of the old Natchez District with the useful features of improved farm implements. He purchased and demonstrated the better-known makes of Northern turning plows, "middle-busters," and sub-soil plows at Ingleside; and he persuaded the influential Jefferson College society to include exhibitions of farm equipment in the programs of its fairs. Like Philips, Affleck made his views on plows and plowing known to a fairly large audience in the South through the press and agricultural periodicals—especially through his agricultural column in the New Orleans *Commercial Times.*[19]

As an agricultural reformer, Affleck was disturbed by the conservatism of Mississippi planters. He always maintained that improper plowing was one of the worst defects of Southern agriculture, and deplored the reluctance of cotton growers to try new methods and implements developed in other sections of the country. Affleck considered the kind of plowing done in the Natchez area harmful as well as ineffective. He wrote that the weight of the plow and the trampling of the teams packed the dirt below plow level into a hard water-resistant "glazed hard-pan . . . under the shallow covering of loose soil . . . so that every rain that falls, runs off, carrying with it large quantities of what little soil is stirred in plowing." Besides promoting erosion, plow penetrations of two or three inches did not loosen the soil suffi-ciently to permit optimum growth of cotton and corn. Yet, according to Affleck, there were Northern implements that could do the job easily and effectively. Sub-soil plows were available which could reach down at least twelve inches below the surface, and cast iron sod-breaking and turning plows made by Ruggles, Nourse and Mason had demonstrated that they could satis-factorily prepare the soil for planting without placing undue strain upon teams of horses or mules.[20]

Philips and Affleck both were hampered in their efforts to persuade Mississippians to purchase superior farm equipment by the activities of a few unscrupulous implement manu-facturers. These industrialists produced inferior implements for the Southern market and sold them as quality merchandise. Planters who paid premium prices for imported Northern plows naturally lost faith in Northern products when the implements did poor work and quickly went to pieces. As cases of this kind were distressingly frequent during the 1840s, the reputation of Northern manufacturers became sadly impaired with Southern buyers. In fact, Southern planters became so suspicious that many refused to buy Northern implements or machines that had been painted. Bitter experience taught them that defects in wood or metal were often concealed beneath beautiful coats of paint.[21]

Because of the uncertain quality of some factory-made implements, Southern purchasers came to rely heavily upon the advice of Philips, Affleck, and other Southern agricultural author-ities. In particular, Philips's reports on tests of plows and other implements were valuable both to the *bona fide* manufacturer and to his customers. In time an unofficial alliance between

reputable plowmakers and Southern agricultural experts gradually succeeded in breaking down some of the reluctance of
cotton growers to purchase late model implements and machinery. Yet progress in this direction was discouragingly slow.
Northern visitors to Southern plantations in the middle 'fifties
were not infrequently surprised by the primitive and ineffective
home-made plows they saw in use.[22] On plantations belonging
to persons interested in agricultural literature and the reform
movement, however, the trend was different. They received the
latest information on new improvements as quickly as did
Northerners, and many used machinery just as good as that on
progressive Northern farms. Indeed, ingenious Southerners succeeded in developing or adapting a family of plows intended
especially for cultivation of cotton and corn, and these innovations were manufactured in vast numbers in a host of local
blacksmith shops.

The principal tasks of cotton growers, aside from planting
and harvesting, always have been keeping the soil loose and
the grass away from the growing plants. In the 'thirties, it will
be remembered, Southwestern cotton planters had depended
largely upon hoes and turning plows for accomplishing these
objectives. In this method of cultivation, plows were of limited
value. Once cotton and corn stalks had begun to grow, plows
could not be run close without cutting roots. Hoe gangs, therefore, had to "scrape" wide strips of ground. As the amount of
labor required to perform this operation was great, cotton
farmers and planters both were eager to find a satisfactory supplement to the hoe.

A crude form of harrow was the first implement used in
Southern cotton fields that gave promise of improving upon the
work of the plow as a cultivating tool. This implement, which
consisted of a wooden framework carrying several rows of short
iron teeth, originally had been employed in preparing the land
for planting. After beds had been thrown up by turning plows,
harrows were dragged across the tops to break up clods and
smooth the surface. Sometimes harrows were used to cover
seeds after they had been dropped into the drills. In the late

'thirties, however, many Mississippi cotton growers found a third use for their harrows. They began to drag them between rows of growing crops to break the crust and remove some of the grass. When used in this way, harrows made additional cultivation with the plow unnecessary. Unlike plows, they could work the entire middle in a single sweep. Thus harrows effected a valuable saving in labor during a critical period of the farm year. Even more important, harrows were shallow-working instruments. They thoroughly pulverized the surface of the earth without disturbing the roots beneath. They could, therefore, be run much closer to the plants than plows. Yet, before full advantage could be taken of the peculiar qualities of harrows, some way of guiding them accurately had to be found. After this problem was solved by adding plow handles, ingenious blacksmiths converted them into several important new farming tools: cultivators, side harrows and double shovels.[23]

In essence, cultivators were refined harrows equipped with plow handles and teeth shaped like small plow shares. Plow-type handles permitted the plowmen to guide cultivators close to the base of rows, and the triangular points broke up the ground very evenly. Some of the cultivators manufactured in Mississippi after 1840 were adjustable in width, and their depth of penetration could be regulated by varying the manner in which the harnesses were attached to the plow beams. In this form, cultivators were versatile machines. They could be used to work all row crops regardless of the width of the middles. Because of these advantages, refined cultivators became standard cotton implements during the 'forties, and have continued ever since to hold that position.[24]

Farmers had only to see a cultivator in operation to appreciate its worth. After witnessing a demonstration of a home-made model in 1842, N. G. North wrote enthusiastically:

> Every day the cultivator is coming more and more into use, as experience tests its advantages; and we do not believe it had ever been laid aside by any one who has once tried it on his farm. Its design is to do *hoe* work with horse power; and it answers the purpose

most admirably—leaving the ground behind it clean, smooth and pulverized in a manner neither the hoe nor the plough can accomplish.[25]

The side-harrow was another member of the harrow family that became popular in Mississippi around 1850. While intended for work quite similar to that performed by the cultivator, it was much simpler and easier to make. In constructing a side-harrow, a crosspiece about three feet long was fastened at an angle of forty-five degrees to the stock of an ordinary plow at the point where the share would normally have been attached. Six teeth made of bar iron were driven through the crosspiece and twisted so as to lie in the same plane as the beam of the plow. When completed, the side-harrow resembled a horse-drawn garden rake. Being light and maneuverable, it served admirably to loosen the earth between the rows of growing crops. Its construction costs were negligible; and its cheapness and effectiveness made it popular with Mississippi cotton growers in the final stages of the depression.[26]

The double shovel, which had been known in the East for at least twenty years, was introduced into Mississippi in 1842. This shallow-running plow was well designed for the cultivation of row crops. Its double plow points with their long wing-like "sweeps" would pulverize a strip of soil two to three feet wide. Like the cultivator, it could be adjusted to run at any desired depth. As all parts of the plow were easy to fabricate, it was produced in quantity by local blacksmiths. After the double shovel gained favor with Southern farmers, Northern implement manufacturers began to produce it for sale through outlets in New Orleans and other Southern market towns. The double shovel increasingly won approval from cotton growers during the closing years of the ante-bellum period, and, like the cultivator, has since remained a part of the planter's stock of farm implements.[27]

Although harrows, side-harrows, cultivators and double shovels all served to loosen the soil and cut down the grass between the rows, none of them touched either the sides or top of rows. "Scraping" off the raised beds upon which cotton or corn was growing was particularly important; for grass growing in the immediate vicinity of the young plants was more harmful to them than grass in the middles. This work was done entirely with hoes in the pre-depression period. In the early 'forties, however, a new instrument for cleaning the sides of rows was put into use by Martin W. Philips. Because it was cheap and efficient the new tool quickly won a place for itself in Mississippi, and from there spread into the other cotton growing states. In Mississippi it was known as a scraper, in other states as a "Mississippi scraper." [28]

Nothing could have been easier for plantation blacksmiths to make than the scraper. In fabricating this tool a flat diamond-shaped piece of iron with one sharp face was fastened to a plow stock at a slight angle to the longitudinal axis of the implement. When in action, the sharp edge of the scraper slid along the sloping side of a row cutting off grass and weeds in the same way that a razor removes whiskers from a face. After the scraper had passed along both sides of a row, nothing but a narrow strip of soil was left on the row top for the hoemen to deal with. This was chopped out by hand. Philips's scraper was a boon to cotton growers everywhere. When used with cultivators and double shovels, it reduced the hoe work required to produce a cotton crop by at least fifty percent.

Scrapers, cultivators, double shovels and other shallow-running implements undoubtedly were important factors in the prosperity enjoyed by the cotton states in the 'fifties. By enabling planters to substitute "horsepower for manpower," they permitted large cotton crops to be grown with much greater economy of labor than ever before. The resulting savings in manpower, however, were seldom used to expand cotton acreages. Picking rather than cultivating was the limiting factor in determining the total size of a crop. What labor-saving devices did for the farmer was to permit him to raise larger crops of corn and other foodstuffs. This in turn lowered the cost of operating a cotton plantation and reduced the cost of producing cotton. Consequently, planters who took advantage of labor-saving equipment were able to farm profitably at times when cotton was selling for low prices.

Although Mississippians were successful in developing or adapting new implements for cultivation, they were less resourceful in devising equipment to aid in planting, harvesting and preparing their crops for market. Mechanical devices for planting corn had long been known to the Northern farmer, but the Southerner was slow to adapt them to his own conditions. This is not to say that cotton producers were unaware of a need for seed planting machines. Isolated attempts were made during the depression era to develop implements that would combine the work of the plowman, sower, and coverer into one operation. In 1842, for example, V. N. T. Moon, a planter of Madison County, reported that he had made and tested "a machine to facilitate planting either cotton, corn, peas, beans, rice, beets, turnips, etc." According to Moon's description, the implement was horse-drawn. It opened a seed furrow, dropped the seed at intervals which could be varied by adjusting the machinery, and then covered the seed and tamped down the earth over them. Moon's planter thus permitted one man and a horse to do the work that had originally been performed by three men and a horse. Despite their labor-saving potential, however, neither Moon's planter nor others of the same order were brought into common use in Mississippi until the late 1850s. Early planters were crudely constructed, and probably did not operate satisfactorily. Furthermore, the need for conserving labor was less urgent during the planting season than it was when the battle against grass was on. This fact alone would have made cotton planters more eager to develop cultivators than they were to adapt Northern corn planters to their requirements.[29]

In the 'forties no one thought of replacing the fingers of human cotton pickers with machinery, although all cotton growers were conscious that their greatest labor-consuming operation was harvesting the staple. The first attempts to devise a picking machine came later—in the second half of the 'fifties.

No new principles for cleaning and packing cotton were developed before 1860. Cotton gins continued in all essential features to be like those made by Eli Whitney in the early 1800s. The materials and standards of workmanship that went into their manufacture were subject to steady improvement, but the volume of cotton they could gin increased very little. Even when gins began to use steam power, the ginstands were not much larger than those driven by horses or mules. Instead, the practice was to gear several ordinary gins to a single source of power.[30]

The situation in regard to cotton presses was much the same as with gins. No new principles were applied to the construction of late model presses. What few improvements were made in them during these years were limited to refining their designs and using better materials in their manufacture.[31] Imported cast iron working parts gradually replaced ones made of wood, a change which permitted the application of greater pressures on the bale. Also the pistons of newer models could be run in and out of the box much faster than those of earlier types. Otherwise, presses remained unchanged. Bagging and ties used in packing bales also changed very little through the years. Cotton ropes and sheeting were used occasionally during the depression, but most planters continued to import hemp rope and bagging from Kentucky for this purpose. In the early 'forties successful experiments were made in which bales were bound with iron hoops instead of rope, and plantation owners considered the new material very satisfactory.[32] New Orleans compress operators, scenting loss of business, however, were able to stave off the general use of hoop iron ties until the closing years of the 'fifties. In short, ginning and baling practices evolved very little during the 1837-1860 period.

During the depression, progress was made in the field of soil conservation which was almost as important to Mississippi agriculturists as the development of improved farm implements. In fact, the savings made possible by use of labor-saving devices would have soon been dissipated by declining soil productivity had Mississippians not succeeded in slowing down the progress of erosion exhaustion of the soil.

The gains made in conservation techniques during the 'forties and 'fifties were based upon a foundation of solid earlier accomplishment. Although soil conservation developments of the 1820s and 1830s were discussed in an earlier chapter, they can be summarized briefly here. Larger planters in Central and South Mississippi had adopted the practice of planting their crops in horizontal rows after the fashion of William Dunbar and Rush Nutt. From the planters of the Gulf Hills they had learned to plow under the dry stalks of cotton and corn instead of burning them in accordance with custom. Furthermore, by the early 'thirties most Southwesterners had begun to plant crops of peas in their cornfields for the double purpose of providing stock feed and enriching the soil. Cotton then was planted upon the land fertilized in this manner. Mississippians thus developed a limited system of crop rotation in which corn, peas and cotton followed one another in that order. Before the panic of 1837 they had begun to employ their surplus cotton seeds as fertilizer. Indeed, by 1837 they had come to grips with problems of conservation so effectively that farmers of later eras were able to add very little to their arsenal of weapons. In general, progress made during the 'forties and 'fifties was the result of improving already known techniques and applying them more widely.

Southwestern cotton growers were more successful in their efforts to check the loss of soil fertility through overcropping than they were in arresting the progress of erosion. Agricultural chemists in the 1830s had revealed that plants extracted elements from the soil. Thus it was known that successive harvests would eventually impoverish the land unless the lost food elements were somehow replaced. Mississippians, in order to minimize the drain upon the soil, adopted the practice of returning as much as possible of the cotton, corn and bean plants to the land. Only cotton fiber, ear corn and fodder were removed from the fields. Cotton seed not used for planting or sold for planting seed were plowed back into the ground, as were the corn stalks and peavines not eaten by the stock. In the late 'thirties and 'forties acreages of corn and peas were increased to approximately half that of cotton, and the three crops were rotated on a three year cycle: two in cotton and one in corn and peas.[33]

In this period a few agricultural reformers tried to improve the common system of rotation by adding small grains and grasses to the sequence. This did not become the custom in Mississippi, however, as grasses were too hard to eradicate from row crops when pasture land was converted to corn and cotton, and small grains impoverished rather than enriched the soil. On the whole, cotton planters of the Southwest were content to alternate cotton with corn and peas.[34]

The cotton-corn-pea sequence did not return enough of the food elements to the soil to prevent a steady deterioration in fertility, but it did slow down the rate of decline. That almost all of the state's agricultural population adopted this system in the 1840s was a great step foward. In later years the deficit in plant food was made up by adding chemicals to the soil. In the 1840s, however, the use of commercial fertilizers seemed wholly impractical.[35] The cost, added to the labor required for successful application, was prohibitive. Similar objections were raised to the use of barnyard manure, which was considered too bulky and too difficult to handle where all hauling was done by wagon. While useful on garden plots and on small areas of land, it was difficult to distribute over fields totalling hundreds of acres.[36] Cotton seed, however, was not subject to this objection.[37] This natural fertilizer could be loaded, transported, and spread easily, and it was a very effective soil builder. Southwesterners knew its value and were in the habit of plowing their surplus seed under to prepare for crops of corn.

Lively discussions were carried on in the agricultural press during the 'forties and 'fifties as to whether cotton seed could be used more profitably as fertilizer or as food for cattle, horses and hogs. In this argument Affleck and Philips held opposite views. Philips maintained that the seed were too valuable as fertilizer to be wasted on animals.[38] Affleck, who thought as an intensive farmer rather than as a cotton planter, advocated feeding the seed to cattle, horses, sheep and hogs, arguing that none of their value was lost in this way if the farmer took pains to save and make use of his barnyard manure.[39] In the main, planters agreed with Philips.

The principal means of fighting soil erosion in these years was the system of horizontal culture introduced into Mississippi by Thomas Jefferson and William Dunbar. This custom of planting crops in rows running horizontally around the contours of hills was almost universal on large plantations of the old

Natchez District and Central Mississippi during the 1830s. From there it spread into the northern portions of the state in the depression period.[40] At the same time, many improvements of importance were made in the technique of "horizontalling."

At the time of the panic, Mississippi cotton growers were laying out their rows by the eye alone. With this method, it was impossible to place rows exactly on a horizontal plane. Water from heavy rains, therefore, would collect in the furrows until it broke through at the lowest point. Gullies then formed where the water overflowed, and the resulting damage to the hillside field was almost as great as if there had been no attempt at horizontal culture. Agriculturists all over the state were dissatisfied with the system, and many of them tried to remedy its defects during the 'forties and 'fifties. At first their efforts were limited to developing means of laying out their rows more accurately.[41] Many crude home-made instruments embodying plumb-bobs or spirit levels were invented for the purpose of determining a horizontal plane. With their aid cotton planters were able to determine almost exact contour lines, mark them with stakes, and direct their plowmen with a fair degree of accuracy. Rows thrown up in this manner held water much more effectively than those surveyed with the eye. Yet they could prevent gullying only in case of light rains. If they filled to overflowing, washing would begin as in the more primitive systems.[42] The next step in improving the horizontal system was to run the rows into drainage ditches.[43] Although ditches permitted rain water to run off without overflowing the tops of the rows, they still did not solve the problem of preventing gullies. Water drained off the fields and rushed down the hillside ditches in large torrents and in time turned the ditches themselves into great gullies. In order to prevent this sort of damage, intelligent planters learned to construct drainage networks so that all hillside ditches would have very gentle slopes. Thus the rate of flow, and consequently the erosive effect, was reduced, and gullies were prevented from forming quickly. At best, however, the system of horizontal rows and guard ditches was able to do no more than to extend the productive life of a hillside field. It could not completely prevent the washing of the soluble Mississippi upland soils. Yet, even with this limitation, the system developed during the 'forties was very valuable to Southwestern farmers. It prolonged the life of productive fields, permitting the owner to obtain a greater return upon his original capital investment before his land became useless for farming purposes.

Several agricultural reformers used a somewhat different method of laying out rows in the 'forties and 'fifties. Instead of plowing along contour lines, as in the system described above, they gave a gentle grade to all their rows. This practice in effect turned the rows themselves into thousands of small drainage ditches which carried off rain water as it fell. There was less tendency to form large gullies than in the guard ditch system, as no great amount of water was channelled into any single ditch. On the other hand, great skill and accurate instruments were necessary to give the rows a uniform predetermined slope. Preparing the land in this manner would tax the ingenuity of a twentieth-century civil engineer using the latest transits. It was accomplished in the late ante-bellum period by only a very small number of experts who were able to employ surveying instruments or modifications of them constructed for this special purpose. Consequently, the effective sloping-row system of culture was not important in the ante-bellum battle against erosion in Mississippi.[44]

The failure of ante-bellum Mississippians to develop methods of effectively checking erosion was no reflection upon their ingenuity or upon their skill as agriculturists. The soils of the hill districts dissolve so readily that no method has ever been devised that completely prevents washing when they are planted in row crops. Only by planting hillsides in trees and grass can the land be saved permanently, and this was not a solution that appealed to a people possessing an economy based upon exportation of staple crops. Under the circumstances, they accomplished as much in soil conservation as any people with equal scientific knowledge could have done. In fact, Southern soil conservationists of the twentieth century have done little more than apply the basic principles laid down by agricultural reformers of the 1840s and 1850s.

CHAPTER X

Prosperity, Technological Progress and Secession

When the 'forties ended, most of the cotton growers of Mississippi had already completed their shift away from single crop farming to a type of agriculture in which cotton was supplemented by livestock and foodstuffs. Furthermore, their adoption of labor-saving devices and new techniques of plant breeding and soil conservation had nearly revolutionized their traditional methods of farming. These changes so substantially reduced the costs of growing cotton that those who accepted them managed to avoid financial disaster even when the price fell below five cents in 1845. In fact, they were able to make modest profits when prices regained the eight cent level late in the decade. Consequently, Mississippians were in position to take advantage of a revival in the international textile trade in the 1850s.

In 1849 textile manufacturers in Britain and New England began to emerge from the doldrums that had gripped them for more than a decade, and the market value of American raw cotton appreciated accordingly. The upward price trend was accentuated that same year by a series of severe late frosts which reduced the Southwestern cotton crop by approximately fifty percent. This combination of strong demand and short supply caused prices at New Orleans to rise from six to eleven cents. As the crop also fell below normal the following season, prices held firm at the new high level through 1850 and 1851. During the next four years the increase in raw cotton production outran the expansion of the textile industry, and the resulting excess of supply over demand drove fiber prices down to an average of nine cents. After 1855, however, consumption again forged ahead of supply, and Mississippi's Upland Middling consistently brought more than ten cents a pound during the latter years of the decade.[1]

The rising price trend brought great prosperity to the agricultural population of Mississippi. Farm income soared to new heights as cotton production and cotton prices rose together. Many small farmers attained the coveted status of slaveholding planters, and not a few large cotton growers were able to amass great wealth.[2] Inasmuch as the greater portion of these cotton profits was devoted to expanding production by purchase of Negroes, the state's slave population grew from 309,878 in 1850 to 436,631 in 1860—an increase of 41 percent.[3] The white population of Mississippi showed no such increase; the number of whites rose only 19.7 percent—from 295,718 to 353,901—during the same ten years.[4] No matter how many thousands of Negroes traders brought to Mississippi from the Atlantic seaboard states, they were never able to supply the avid demand.[5] Hence the average price of an able-bodied field hand on the New Orleans market rose from $700 in 1845, to $1,100 in 1850, to $1,800 in 1860. This dramatic increase was a reliable gauge of the prosperity of Mississippi planters and would-be planters during these years of expansion.[6]

Nevertheless, not all of Mississippi's cotton profits were devoted to increasing the size of the principal crop. During the 'fifties many planters, merchants and bankers invested thousands of dollars in the construction of a network of railroads crossing the state. Growth was very rapid in this field after the depression. Only seventy-five miles of track were actually in use in 1850, but the total of active mileage reached 872.3 in 1860.[7] By that time eight different companies were operating railroads, and their combined investment in roadbeds and rolling stock exceeded twenty-four million dollars, virtually all of which had been contributed by Mississippi investors.[8] As the state's cotton growers showed an inclination to ship their cotton by rail when ever possible, the earnings of the railways swelled in a manner very pleasing to their stockholders. The experience of one line was typical. The Mobile and Ohio, which went into operation in Mississippi in 1857, transported 107,450 bales of cotton during the 1857-58 season, 137,430 in 1858-59, and 223,890 bales in 1859-60. Its profits for the three years were $617,501.87, $773,-179.18, and $1,200,108.61.[9]

Local capital in these years also found its way into a small but expanding textile industry. The number of factories producing cotton and woolen goods increased from two to six between 1850 and 1860; the number of spindles from less than 1,500 to more than 8,000, the number of mill workers from 58 to 717, and the money invested in plants and equipment from less than $50,000 to a total of $345,000.[10]

Still other cotton dollars were spent in elevating the living standard of the planter class. Many prosperous cotton growers sent their sons to the University of Mississippi and a few to Harvard, Princeton, or the University of Virginia. Some took their families to resorts in both the North and the South to escape the summer heat. To all of them, luxuries of the 'forties became common necessities of the 'fifties.

Changes in living standards were especially evident in the residences of the wealthier planter class. New houses were erected in town and country to replace the common log cabins and frame cottages of the depression period. Although unpretentious when compared with the famous mansions of antebellum Natchez millionaires, these new homes were much handsomer and more substantial than the pioneer structures they superseded. In architecture they all were very much alike, rectangular or T-shaped, with wide verandas of one or two stories. They were heated from wood-burning fireplaces set into chimneys located at each end of the house. Because of the ever-present danger of fire, their kitchens characteristically were placed in separate buildings situated at some distance from the "Big House." [11]

Despite its material prosperity, the last decade of the antebellum era in Mississippi was not marked by agricultural progress such as had characterized the preceding twelve years.

Because most Southwestern cotton growers were satisfied with the adjustments already made in their system of farming, they adopted few innovations after 1850. Not all progress ceased, of course, but such advances as were made followed lines already plotted during the depression period. In short, Mississippians were generally content in the 'fifties to reap the golden harvest sown in the late 'thirties and 'forties.

There was even a reverse swing of the pendulum in some aspects of Southwestern agriculture during the ten years preceding the Civil War. The resolution of many cotton growers to achieve self-sufficiency was weakened by swelling profits derived from the sale of cotton. Having lost some of their aversion for importing food stuffs, livestock, and manufactured goods from the Upper Mississippi Valley once they had cash in their pockets, both planters and farmers tended to revert to habits of yesteryear. They imported articles which they could have produced themselves, and gave less attention to growing other crops than cotton. At the same time larger planters were losing their enthusiasm for raising fine livestock, and their interest in building pastures was on the wane. Indeed, the ideal of economic self-sufficiency that had been prominent in Mississippi during the 'forties seems to have lost most of its force by 1860. Planters then had become concerned mainly with producing the maximum number of cotton bales. Hence major agricultural developments in Mississippi in the 'fifties were limited almost entirely to improvements directly affecting the staple crop itself. Farm implements were further improved, soil conservation practices became more common, superior breeds of cotton became increasingly popular, commercial fertilizers were put to the test, and planters for the first time began to wrestle with the insect problem.

Improvement in farm implements was particularly noteworthy in this decade. Several models of seed planters were perfected; implements used in cultivating the growing crops multiplied in numbers and types; steel sod-breaking and turning plows slowly replaced those made of cast iron; and a few attempts were made to devise machines for gathering cotton. Now that they could afford the cost, large planters put steam engines to use in driving gins, presses, saws, and mills of many kinds; and agricultural papers occasionally discussed the possibility of making steam plows. Manpower, in fact, was slowly giving way to horsepower and steam power in many of Mississippi's farming operations toward the close of the ante-bellum era.

Many of the improvements made in farm implements during this decade were due to the ingenuity of Southerners. Northern mechanics and plow manufacturers continued as before to excel in the design and manufacture of turning and breaking plows, horse rakes, reapers, mowers, and other implements and machines useful to the Northern grain farmer. The development and manufacture of specialized equipment for cotton plantations were left largely to Southerners themselves, and craftsmen below the Mason and Dixon line were equal to the task. Cultivators, double shovels, sweeps, bull tongues, harrows, cotton scrapers, and many other tools used in the shallow cultivation of row crops were fabricated in shops throughout the Lower South. While few subsoil and sod-breaking plows were produced in the cotton states during these years, light turning plows were manufactured in considerable numbers by local plow makers and plantation blacksmiths. The situation was much the same with regard to plantation machinery. Almost all cotton presses were built on the spot where put to use, although their more complicated metal parts were purchased from foundries in Kentucky and Ohio. Southern ginwrights also produced large quantities of gins embodying moving parts of Northern manufacture, and these locally made machines were said to be fully as efficient as those fabricated by Carver and other New England manufacturers.[12]

Mississippians did not lag far behind in making such farm equipment as gins, presses, plows, and cultivators. Their farm implement industry, it is true, was operated, with one exception, on a small shop basis; but its output was sufficient to supply the local market with most of the implements used in planting and cultivating corn and cotton. Furthermore, a few ingenious cotton growers and craftsmen were successful in developing a number of labor-saving devices that won favor with agriculturists of the state.

In 1855 George W. N. Yost, an overseer from the neighborhood of Port Gibson, invented and patented a combination
cotton scraper and turning plow intended to combine two separate operations in cotton cultivation, thus saving the labor of a
team and a field hand. When Yost's implement was drawn along
between the cotton rows, its scraper blade cleaned the grass from
the side of one of them. Simultaneously a turning plow point
behind the scraper opened a water furrow alongside the row and
threw fresh dirt upon the side of the row which the scraper had
just cleaned. The "Plow and Scraper" also was useful in preparing ground for planting. After the earth had been broken up
by two-horse sod-breaking plows, the Yost implement could be
used to throw up seed beds. In doing this job, the Yost had a
definite advantage over conventional turning plows: its scraper
blade could be adjusted to smooth off the top of the bed behind
the turning plow point.[13]

When Yost gave the first public demonstration of his new
implement, his planter audience was much impressed. Among
those present was Thomas S. Dabney, a wealthy Hinds County
slaveowner, who ordered twenty-four of the Yost plows. Other
planters bought them almost as eagerly as Dabney. In fact, more
than one hundred orders were placed that day with the Hinds
County distributor, even though the price was set at the relatively high level of ten dollars per plow. Moreover, Dabney and
thirteen of his fellow planters volunteered written testimonials
of the value of the Yost implement.[14]

The Yost Plow and Scraper gained in popularity during the
next five years, and it was introduced into all cotton producing
sections of the state. Garland D. Harmon, a literate, scientific,
and highly skilled professional overseer, was especially enthusiastic about its merits, and he warmly recommended it to cotton
planters everywhere. In a letter to the *Southern Cultivator* he
wrote:

> The cotton planters of the South have long wished
> for an implement that would scrape cotton cleanly and
> nicely, and at the same time clean and pulverize the
> row, whether three feet or six feet wide; thus doing the

work at a single round that has always taken two rounds to perform. That implement was never found until Yost invented the "Plow and Scraper" combined.[15]

Planters usually agreed with Harmon after they had given it a trial. Samuel Redus, agent for the Mississippi Manufacturing Company, noticed that in every section he visited cotton growers were using Yost's plows; and he reported that "they all, without a single exception, pronounced them the best cotton plow ever invented." [16]

Yost, having neither capital nor facilities for manufacturing his plow, was compelled to follow the usual procedure of inventors of farm equipment during the late ante-bellum period. After obtaining a patent, he sold the privilege of manufacturing and distributing the Plow and Scraper to individuals whom he authorized to operate within a given district. In one instance N. B. Ward of Hinds County obtained the right to manufacture and sell the Yost plow in four counties at a cost of only five hundred dollars. Ward then engaged a firm in Pittsburgh to produce the implement for him. They were made in four sizes so as to fit rows of varying widths, and Ward sold them at prices ranging from ten to twelve dollars each. Yost himself left Mississippi in 1856 and subsequently entered the plow manufacturing business in the North.[17]

Dr. A. W. Washburn, planter, cotton breeder, and horticulturist of Yazoo County, was also an inventor of some prominence. In 1858, Washburn patented two farm implements and placed them on the Southern market. One of these was a mechanical seed planter which opened a seed furrow to any desired depth, dropped seed at any interval selected by the operator, and then covered over the furrow, all in one operation. This planter was drawn by a single horse or mule and directed by a single field hand. The seeds to be planted were put into a barrel rotated slowly by means of a chain connected to a large roller serving in place of wheels. When the planter was in motion, seed passed through perforations in the barrel into a spout. From the spout they fell into the furrow opened by a plowpoint attached to the front of the implement. A roller on the rear of the planter

covered the seeds. A movable metal band around the waist of
the barrel regulated the number of seed dropped by the planter
by increasing or decreasing the number of open perforations.
Washburn claimed that his planter would perform equally well
with either cotton seed or corn and that it could be used to drop
guano or dry manure just as well. The Washburn planter sold
for fifty dollars delivered at any point in the Southern states.[18]

Washburn gave his second implement the name "Hiller and
Scraper." This odd and cumbersome device was drawn by a
double team, and its multiple purpose was to open a furrow
along each side of a row, throw dirt upon both sides of the row,
and scrape the row clean of grass. Its carriage consisted of a
horizontal wooden framework mounted upon four large wooden
wheels. Two iron scraper blades and two turning plow points
were attached beneath the body of the machine. As the Hiller
and Scraper was pulled along straddling a row, the scrapers in
front cut the grass from both sides. The trailing plow points
threw loose dirt upon the freshly cleaned row. Washburn's
machine performed satisfactorily on dry ground so long as the
cotton plants were short. After they grew a foot high, the heavy
Hiller and Scraper could not pass over a row without crushing
them. Thus the implement's usefulness was restricted to the
first and second workings of a crop.[19]

Washburn fabricated planters and Hillers and Scrapers in a
small shop in Yazoo City to order only, and no records have
survived to indicate the extent of his business. As he advertised
in papers with a Southwide circulation for several years, he must
have sold enough machines to pay promotional expenses.[20]

Several other multiple plows were put into use in Mississippi
in the late 'fifties in competition with Yost and Washburn
devices, and sooner or later each of them was given a test by
Martin W. Philips. In 1859 he announced in the *Southern Cultivator* that he recently completed a series of tests of two Mis-
sissippi-invented and manufactured double scrapers.[21] One of
these, a two-horse double scraper patented by J. G. Winger,
had worked well on Philips's comparatively level but well
drained land. Like Washburn's Hiller and Scraper, the Winger

machine straddled a row and scraped both sides at once. G. D. Harmon, who also was familiar with the Winger scraper, was less favorably impressed with it than Philips. When he tried to employ it on a hilly Hinds County plantation, he learned that any obstruction in the way of one wheel would cause the machine to pivot and cut into the row. On perfectly clean, level land, however, he found that it performed quite satisfactorily.[22] The second double scraper that Philips had tested, an implement invented by Patrick Sharkey of Hinds County, was meant to be an improvement upon the earlier Winger device.[23] Its points were set one in advance of the other rather than side by side in the hope that this arrangement would prevent the machine from gashing the row when one wheel encountered an obstacle. Philips found it to be useful, but Harmon was less charitable in this case also. "The Winger" [and others like it] he remarked upon one occasion, "will do where there is *nothing* to do. If you want the best Scraper on the top side of the globe, get Yost's Plow and Scraper."[24]

Many planters agreed with Philips rather than with Harmon about the value of the Winger scraper. In 1858 seventeen prominent Central Mississippi planters, including John Hebron and Pearce Noland of Warren County, certified that they had used it the preceding spring and had "found it the most labor-saving and neatest working implement we ever used, or have seen in a cotton field."[25] The implement was placed on sale during the late 'fifties for twenty dollars each at Greenville, Natchez, Vicksburg, Port Washington, and Carolina Landing, in Mississippi, and at Lake Providence, Goodrich's Landing, and Milliken's Bend, in Louisiana.[26]

The Stuart Double Plow and Double Scraper was one of the most interesting devices put on the Southern market in the years immediately preceding the Civil War. It was invented by a Negro blacksmith belonging to Oscar J. E. Stuart of Holmesville, Mississippi. D. H. Quin, who had used the Yost scraper, the Taylor (or simple Mississippi) scraper, and the Baggett & Marshall Plow and Scraper on his plantation, regarded this slave-invented machine as decidedly superior to the other three:

With it, [he said] one hand and two horses will do double the work in good ground of the Baggett & Marshall Scraper, and the work of four hands, four horses, two common baring ploughs, and two common Scrapers. . . . By taking off the Scraper and reversing the ploughs, it may also be used for hilling a row on both sides at once, of either cotton or corn, and by causing both horses to walk in the same water furrow, it will hill a row of either upon the right and left without reversing the ploughs.[27]

Ex-Governor Albert G. Brown, of Hinds County, was equally impressed with the Stuart machine. In 1859 he informed Stuart:

The Taylor was a great improvement on the Hoe, the Yost patent was a great improvement on that; but your "Double Plow and Scraper" goes a long way ahead of both. When it shall be made by machinery instead of being hammered out in a country smith shop, it will, in my judgment, be the very best agricultural implement ever offered to the cotton planter. I am glad to know that your implement is the invention of a negro slave—thus giving the lie to the abolition cry that slavery dwarfs the mind of the negro.[28]

Stuart soon learned that securing a patent for an implement invented by a slave was not easy. Under Mississippi law a slave was not permitted to hold property; and his owner could not truthfully certify that the slave's invention had been the work of the would-be patentee as the Patent Office required. Although a lawyer, Stuart could find no way out of this legal impasse. He therefore tried to obtain aid and advice from friends in the national capital. Fearing that a Northern-born Commissioner of Patents might be hostile to his project, the troubled Mississippian appealed directly to Jacob Thompson, the Secretary of Interior, who, as a friend and fellow citizen of Mississippi, "would be exempt from all the prejudices which might cloud the under-

standing of a man from a different latitude." [29] Upon learning
that the Secretary could not solve his problem, Stuart next
sought the assistance of his senator. He wrote a letter to Senator
John A. Quitman, of Mississippi, explaining the situation in detail
and asking him to sponsor a special bill extending the protection
of the patent laws to Stuart's implement. Quitman no doubt was
properly sympathetic, but he failed to obtain the desired legislation. Since two such powerful friends as these had failed him,
Stuart gave up hope of patenting his slave invention and resolved
to produce and sell it without the usual legal safeguards. He
manufactured and distributed a number in Mississippi prior to
secession. As planters bought them readily, he made plans to
build a steam implement factory for turning them out on a large
scale. But the coming of the war destroyed that project.[30]

Farm implements were manufactured in Mississippi on a
true factory basis in only one establishment during the ante-
bellum period. That was the Southern Implement Manufacturing
Company located at Jackson, a concern owned by Martin W.
Philips, his brother, Z. A. Philips, and his son-in-law, Robert
Kells. This business enterprise was the outgrowth of Philips's
life-long interest in improving farm equipment. In 1856 the
Mississippi experimenter had received a cast steel Brinley
turning plow as a gift from the Louisville, Kentucky, plow
designer and manufacturer, Thomas E. C. Brinley, who hoped
to obtain Philips's endorsement for his implement. This plow had
been placed on the market in the Upper Mississippi Valley a
few years previously, and it had already won a reputation for
fine quality in that section of the country. Brinley, however,
wished to extend his business into the cotton growing states,
and a recommendation from Philips was sure to bring him many
orders from the Lower South. Philips was glad to conduct tests
with the Brinley implement. He and his overseer, G. D. Harmon,
experimented with it for a time, eventually becoming convinced
that the cast steel Kentucky plow was the most effective in their
collection of more than one hundred makes and types. Philips
in particular was so favorably impressed by the performance of
this Brinley implement that he determined to manufacture and

sell it himself. Having full confidence in Philips's judgement where farm equipment was concerned, Z. A. Philips and Robert Kells were easily persuaded to join in the venture. The three men organized a company in 1857 and erected a factory in Jackson.[31]

Philips & Kells experienced no difficulty in securing permission to manufacture the Brinley line of plows. Martin W. Philips invited Brinley to leave Louisville and come to Mississippi to assume the management of the newly organized Southern Implement Company. Brinley accepted this offer, and in return gave the company the right to manufacture and distribute his plows.[32]

Brinley supervised the construction of the implement factory and probably was responsible for its design. The work on the buildings and the installation of machinery progressed slowly but steadily under his direction, and the plant was ready to begin operation early in 1859. At that time it consisted of a three story brick building measuring forty by one hundred feet. Two other smaller buildings stood adjacent to the main structure. One other with dimensions of forty by forty feet was located in the rear. The factory was powered with a steam engine and equipped with a wide variety of woodworking machines, including a planer, sharpers, saws and turning and morticing machines. Thus the factory was prepared to produce all wooden parts used in the manufacture of wagons, plows and other types of farm equipment. Lacking a foundry, however, the company was compelled to import all iron and steel castings. Philips and his partners had plans to remedy this deficiency, but the outbreak of war upset that design.[33]

By the summer of 1860 the Southern Implement Manufacturing Company was in full operation, producing plows, scrapers, sweeps, harrows, wagons, carts, and wheelbarrows. The company made implements of both iron and steel and sold its iron products at somewhat lower prices than similar ones of steel. It also manufactured several types of wagons, which ranged in size from one to six-horse units, and were designed to be drawn by either horses or oxen. The axles of these wagons were usually made of wood, but customers could obtain iron axles if they preferred.[34]

The Jackson implement and wagon factory manufactured and distributed a large number of farm implements and wagons in 1859 and 1860 through outlets in Jackson and Vicksburg. Their products were priced low enough to compete with similar articles produced for the Southern trade by factories in Louisville, Kentucky. But from the start Philips and Kells were afflicted by unethical competition from local Mississippi plowmakers who copied their designs and sold the fakes more cheaply than the implements made in Jackson. In some instances the counterfeiters even sold their wares with the Philips & Kells initials painted on them. How the owners of the Jackson company coped with this problem is unknown; for no information about the operations of the plant has come to light for the period after 1860—presumably because all the company's records perished in 1863 when the factory was burned by federal troops.[35]

The Brinley cast steel turning and sod-breaking plows were the most popular items produced by Philips & Kells. In appearance these implements were very like that of horse-drawn plows of the mid-twentieth century. Their beams and handles were made of wood, but all working parts were cast steel. In performance these Mississippi-made plows were extraordinarily good. In a test conducted by disinterested parties, one of Philips & Kells' cast steel plows opened a furrow nine inches deep and ten inches wide, while its nearest competitor, a Northern Hall & Spere, was able to cut no more than a seven by seven and a half inch slice of earth. In this test, the Brinley registered a draft of only 350 inches on a dynamometer while the Hall & Spere pulled 623 inches.[36] G. D. Harmon, who was an authority on plows second only to Philips himself, reported in 1859—after having left Philips's employ—that he could always obtain a ten inch furrow with either a Brinley sod-breaking or turning plow.[37] By 1860 the Brinleys made by Philips & Kells were selling so rapidly that the factory was unable to keep pace with the demand.[38]

With the wide-spread adoption of the cast-steel Brinley plows, plowing in Mississippi reached a high degree of efficiency. Since the 1850s, materials used in the manufacture of turning and breaking plows have been greatly improved by the development of alloy steels, but few later models have surpassed the record-making performance of the Brinley. Its ten by ten inch furrow stands as an achievement worthy of respect even a century later.

The improvement in design and manufacture of farm implements was matched during the 'fifties by the development of accurate instruments used in laying out horizontal rows and graded drainage ditches. The crude homemade wooden plumb bob devices commonly employed by Southern planters in the 1840s were rapidly replaced on larger plantations by newly invented instruments, patterned upon the surveyor's transit, which employed precision adjustments, spirit levels, and very accurate workmanship. One of the best of these new devices was invented in the late 'fifties by Joseph Gray, of Raymond, Mississippi. Gray, like many other progressive cotton growers of Central Mississippi, had for many years battled soil erosion with only partial success. Like them, he discovered that horizontal rows laid out by eye or with the aid of crude plumb bob devices were not effective enough to arrest the progressive washing of the soil. By 1857 the situation in Gray's home county, according to the Raymond *Hinds County Gazette,* was growing critical. The editor of the *Gazette* said:

> As the citizens of all this region are abundantly aware, the greatest difficulty with which our farmers and planters have to contend is the constant "washing" to which their lands are subject from the moment they are put into cultivation. Scarcely a plantation in Hinds County, probably, is entirely exempt from this annoying and perplexing fault; and, certainly, we have seen immense fields so completely riddled with "washes" as to be abandoned as utterly worthless. In many instances, even the most careful and scientific management has failed to secure broad acres from this destruction— a destruction not unlike that which awaits the sandbar when its front is presented to the dashing floods of the great Father of Waters.[39]

Gray, unlike his neighbors, worked on the problem of erosion until he found a solution. After years of experiments, he finally became convinced that a system of graded drainage ditches connected to accurately laid out horizontal rows could stop the formation of gullies. On applying this system to his land, he found that the problem of giving the proper grade to the ditches was most troublesome. None of the instruments known to him provided the results he desired; so he devised one of his own. In essence, it was a surveyor's transit—minus the telescope—that had been adapted to perform a specific task. He deposited a drawing and a model of his "Grade and Horizontal Level" with the U. S. Patent Office in 1857, and was granted a patent in the following year.[40]

Gray's plantation became the best advertisement of his instrument. After he had given his ditches the desired grade, had laid out his rows along contours, and had planted his hillsides in "Rescue" grass, he ceased to be troubled by erosion. G. D. Harmon visited the Gray plantation in 1858 and reported that— to his amazement—no "washes" were to be seen anywhere upon its rolling hills.[41]

Philips was much interested in instruments of this type, and he invited Gray to demonstrate his level at Log Hall. Gray accepted the invitation gladly. On July 23, 1858, he surveyed a row 180 yards long with a fall of three inches to twelve feet, and the slope of the row when checked with a transit was found to be extremely accurate. The demonstration was so conclusive that many of the planters witnessing the test placed orders for the Gray level.[42]

Wishing to give his instrument a chance to prove its potentialities, Gray sent a model in 1859 to the pioneer agricultural school at Montpelier, Georgia—the Montpelier Farm School, directed by Carlisle P. B. Martin. Martin employed the Gray level in his surveying classes the following season, and gave it his unqualified endorsement. "I am now able to report *in extenso,* having made the most thorough trial of it on all kinds of ground and in various sorts of work; such as running horizontal rows and grading ditches," he said in a letter to the *Southern Cultivator.* "In every trial it has performed to my entire satisfaction, and

I can say that I have seen no instrument constructed for the purpose at all equal to it." The Gray level thus was proven to be a practical instrument. Not only could skilled surveyors lay out accurate rows and ditches with it, but so could untrained boys after a little instruction.[43]

The Gray level was selling briskly in all of the cotton states in 1860, but the outbreak of the Civil War turned the attention of Southern farmers away from battles against soil erosion to struggles of a more bloody nature. Consequently, instruments like Gray's were not put to work on a large scale until many another decade had passed.

Mississippi cotton growers made numerous tests of commercial fertilizers during the 'fifties, but found none to meet the requirements of extensive cotton cultivation. The science of soil chemistry was far enough advanced by 1850 to enable planters to know that fertilizers of some types were needed to prevent deterioration of their lands, and chemical analyses of cotton and corn had determined the elements that these plants removed from the soil.[44] Furthermore, analyses of the soil itself had demonstrated that different lands were likely to have different deficiencies in plant food. "There may be lands rich in minerals and poor in vegetable substance," explained Dr. H. D. Webb, of Kemper County in 1850.[45] "To these it would be folly to supply what is already superabundant." What such lands need, he continued, is that carbonic acid which is supplied by compost manures. For best results, he concluded, chemical tests should be made to determine exactly what the soil lacked; then the proper chemicals or organic matter could be added profitably.[46]

Lime, plaster of Paris, and several phosphate compounds could be purchased in Mississippi during the 'fifties, and farmers occasionally used these fertilizers to advantage on patches of clover and other small sized crops. They were too expensive, however, to be used upon extensive cotton fields. The same was true of guano, which was being imported from South America in vast quantities during the last decade of the ante-bellum period. Many planters who tried this bird manure realized greatly increased crop yields, but the increase was not sufficiently large to offset the expense. The following experiment reported by Eli J. Capell, a scientific planter of Amite County, was typical:

> I experimented last year with 300 pounds on a piece of old clay land that had been cleared about thirty years with the following result: I measured one acre and divided it into two cuts and had it plowed into beds four feet apart. On one of the half acres I put 151 pounds of Guano, sprinkled in the water furrow, and had two furrows turned back on it. On the 26th April [I] opened the ridge and planted corn, and thinned it to two stalks in each hill and three feet apart (too thick) and cultivated as usual. Upon gathering the crop 1 found that the half acre with Guano made fifteen bushels of shelled corn, and the half acre without Guano, four bushels, one peck and two quarts.[47]

A similar experiment that Capell conducted with cotton produced 619 pounds of seed cotton on a measured half acre fertilized with guano and only 194 pounds without it. Great though this increase was, Capell concluded that the use of guano was unprofitable. "I am delighted with it as a manure," he wrote, "but do not think it will pay at present prices. We must depend upon homemade manure, rest our land and shade it, and keep it from being trampled by stock."[48]

Because many other experimenters ultimately arrived at the same conclusion, ante-bellum Mississippians did not adopt commercial chemical fertilizers and imported manures. Instead, they continued to depend primarily upon legumes, cotton seed, and corn and cotton stalks for fertilizers. In fact, very few even bothered to spread manure from lots and barnyards upon their lands.

Frightened by the increasing severity of damage to their crops by "caterpillars and boll worms," Mississippians in the 'fifties began to seek means of combatting these pests.[49] All their efforts, however, were completely ineffective, and many of them were ludicrous when viewed through twentieth century eyes. In some instances, planters hung white flags upon poles planted throughout their cotton fields in hopes that insects would lay their eggs upon the cloth instead of the cotton plants. In others, harassed cotton growers placed pans of molasses on the ground between the rows so that the insects would be attracted by the smell and trapped by the sticky fluid. Still others had slaves carry blazing torches through the fields at night to lure moths to their deaths in the fire, while a few tried to accomplish the same object by lighting many bonfires at strategic locations about their plantations. Some planters placed their faith in the beneficent effects of noise. They attacked the swarms of flying insects infesting their cotton fields by beating upon drums, pots, bells and other utensils. "Hebron," of Lawrence County, was even more ingenious. He turned pigs into his cotton to eat army worms and caterpillars. Just how he was able to train his animals to so restrictive a diet when afield, he never made quite clear, but he was confident that he had hit upon the solution to this pressing problem. A great majority of Mississippi farmers and planters were less gifted with imagination, and they contented themselves with resorting to curses or prayers as their natures dictated.[50]

One development of Mississippi society during the 'fifties was a clearly defined trend toward organization on the political, social and economic levels. In some ways this tendency of the farming community toward cooperation was reminiscent of the movement for agricultural improvement of the late 'thirties and early 'forties; in others, it was quite different. Where the formation of agricultural societies in the depression era had been prompted by financial distress, exactly the reverse was true during the late 'fifties. Extremely prosperous cotton farmers and planters were responsible for reviving the old state and county agricultural societies and for forming many new farm organizations between 1855 and 1860. These cotton growers were concerned primarily with improving their cotton production rather than with achieving balance in their agriculture. Their state and county fairs emphasized the staple instead of livestock, and more visitors were attracted to the fairs by their entertainment facilities than by their agricultural exhibits. Perhaps even more significant, the new farm organizations showed signs of having lost the non-political character which had distinguished their predecessors. The earlier leaders in agricultural improvement had worked together to achieve a common objective without regard for political affiliations, but, in the late 'fifties, they were replaced by agricultural politicians closely linked with the secessionist wing of the Mississippi Democratic party. In short, the agricultural organizations of the late 1850s were seeking different goals from those of the depression period. Instead of aiming at economic self-sufficiency, more and more farm leaders were turning to political independence as the solution for their problems.

The change in the basic nature of Mississippi farm organizations was due partly to the belated realization that politics could have an influence upon economic developments. Calhoun and other spokesmen of the Old South had made it clear to their constituents that tariffs and other works of the government in Washington had accomplished miracles on behalf of the industrial sections of the nation—at the expense of the agricultural South and West. Consequently, many Southerners began to wonder whether their state governments could perform similar services for agriculture. At first, their demands upon state legislatures were modest, but as time wore on they became more insistent. What was true of the South as a whole in this regard was true also of Mississippi. In the late 'forties and early 'fifties enough of the people of the state were interested in surveys of the geological and agricultural resources to inspire the legislature to authorize three such investigations. Similar pressure from farmers and planters brought about the establishment of public facilities at the University at Oxford for making chemical analyses of soil, farm products, and fertilizers. In addition, a Democratic legislature that was anxious to conciliate the farming interests of the state passed an act in 1857 creating an agricultural bureau as an adjunct of the executive branch of the state government. Obviously, popular demand for participation of the government in agricultural matters had brought agriculture and farm organizations into the political arena.

In 1850 and 1851 a small number of local agricultural associations dating from the 1840s showed signs of revival, and this activity encouraged a few of the old farm leaders to begin work toward establishing a statewide organization once more.[51] Their efforts were rewarded with but small results until Martin W. Philips issued an appeal for the formation of such an organization in the fall of 1853. Philips proposed that "those of the right kind of spirit . . . for the good of the cause and for the welfare of our State" convene at Jackson at the time of the next legislative session.[52] Acting for a body of these agricultural reformers, Philips in November suggested that the convention be held on January 17, 1854, as many members of the state Masonic order would be in the city as well as the legislators.[53]

The agricultural convention, however, assembled on February 3rd instead of the date named by Philips. That day a "respectable number of the friends of Agricultural and Mechanical improvement" met in the chamber of the House of Representatives. The delegates—mostly self-appointed—adopted a constitution written by Philips and elected a slate of officers. Governor William McWillie was named president and B. L. C. Wailes, Eli T. Montgomery, W. R. Cannon, Jacob Thompson, Henry W. Vick, J. S. Yerger, J. M. Hands, C. L. Thomas, and T. P. Stubbs were elected vice presidents. Richard Griffith became recording secretary; F. B. Hunt, treasurer; and M. W. Philips, corresponding secretary. S. C. Farrar, J. B. Peyton, N. M. Taylor, Robert Shotwell, Benjamin Ricks, John Hebron and John Robinson were appointed to the executive committee. Beyond completing the process of organization, however, they did nothing more at the time.[54]

It will be noted that the leaders of this state agricultural society were men who had been active in the agricultural reform movement of the depression period. Wailes, Philips, Hebron, Vick, Montgomery, Peyton, Ricks, and Farrar, for example, had worked together over the years to promote the cause, even though they did not all share the same political convictions. Vick, Hebron, Wailes and Farrar were Whigs, while Philips and Montgomery were Democrats. Clearly the new farm organization was in the non-political tradition of the reform movement of the 'forties.

The purpose of the new society was also in keeping with that of the earlier organization. It was intended to promote state fairs, and it gave lip service, at least, to the ideal of a diversified agriculture. Philips, however, indicated that a change in goal was in prospect. In an open letter to the planters of the state, he suggested on March 18, 1854, that the society shift its emphasis from diversification of crops to improving the cultivation and increasing the production of the principal staple crop. In his opinion, the state organization's fairs in future should be used to promote experimentation with new methods of farming. "We should encourage a better system [of cultivation of cotton and corn]," he wrote, "and offer premiums for what will best draw out that system." [55]

Despite energetic efforts by Philips, Wailes, and a few other agricultural enthusiasts, the new state organization failed to accomplish anything of note. Its members seldom attended meetings, and nothing came of the attempt to sponsor or hold state fairs. Although this society clung to a precarious existence for several years, its record made it clear that the old ideal of agricultural self-sufficiency was dead and that most of the leaders and workers of the earlier period had lost interest in diversified farming by 1855.[56]

This does not mean that Mississippi cotton growers were blind to the value of agricultural cooperation. On the contrary, they wanted a state-wide farm organization, but the kind they wanted was quite different from that of the depression period. The changed climate of opinion in Mississippi agricultural circles was recognized and described by the editor of the Aberdeen *Independent* as early as 1852. While urging the farmers of Monroe County to attend the organizational meeting of a new agricultural society, he remarked:

As a class they [cotton growers] have general interests and duties they should not fail to attend to. These the Society is intended to promote. The farmers of the South lose vast amounts by failing to attend their interests as a class—by want of concert and organization. The cotton broker, the cotton spinner, the cotton speculator have a concert of action among themselves and conspire against the interests of the cotton producers, but the cotton producers do nothing to counteract their organization. They are *disorganized* while their enemies are *organized,* and the latter are consequently enabled to fleece the farmer to their hearts' content.[57]

Obviously, the editor of the *Independent* wanted an organization that could bring pressure upon the state and federal governments to enact legislation favorable to the cotton grower, and one that could aid these planters in getting better prices for their products. He had little or no concern for crop diversification or soil conservation.

The state legislature in 1857 demonstrated that the views of the *Independent* were representative of a large segment of public opinion. With remarkably little discussion—either in the newspapers or on the floor of the Senate or House—a bill introduced by Thomas J. Hudson, a prominent Democrat from Marshall County, was passed and signed by the governor. Hudson's bill created a Mississippi Agricultural Bureau that was meant to do for the agricultural community of the state what the "National Bureau" was doing for the farmers of the nation. Its purposes, as asserted by the law, were to gather and disseminate information and statistics about all aspects of farming, to promote the raising of fine breeds of animals and plants, and to establish a chemical laboratory for analysis of soils, manures, and minerals. Even more important, it was intended to organize a state-wide network of county agricultural societies subordinated to the Bureau.[58]

The structure of the Mississippi Agricultural Bureau was quite simple. It had a president, a secretary, and an executive committee. The executive committee was composed of persons chosen by the state legislators—one for each judicial district— and the governor of the state, who sat as an *ex officio* member.

The presidency of the Bureau was largely an honorary office. The only duty of the president was to serve as chairman of executive committee sessions. The secretary was given the task of carrying out the directives of the committee, and he, therefore, was the only official of real importance in the Bureau.[59]

From 1857 to 1860 the legislature provided an annual fund of $1000 to finance the operations of the Bureau. Half of this sum was used to hold a state fair, and the remainder to defray the incidental cost of the Bureau, which in practice was the salary of the secretary. Additional funds were made available to the Bureau from the state treasury for subsidizing local agricultural organizations associated with the state agency. According to the law, the secretary of the Bureau was to provide each of these societies with $200 when it had raised $300 itself. This subsidy was to be used in holding local fairs.[60]

At the first meeting of the Bureau, Thomas J. Hudson was elected president, in recognition of his services in promoting its creation, and John J. Williams, editor of the *Planter and Mechanic,* was appointed secretary. The executive committee at that date was composed of Governor McWillie, John J. Pettus, J. B. Herring, W. P. Anderson, T. J. Hudson, and E. Brown—all leading Democrats. Governor McWillie and W. P. Anderson were placed in charge of arrangements for the first state fair to be held under the auspices of the Bureau.[61]

After assuming office as permanent secretary, Williams threw himself into the task of building up a network of local agricultural societies. He invited all existing associations to affiliate with the Bureau, and when applications for membership came in slowly, he went in person to persuade them to join hands with the state agency.[62] He addressed gatherings in all sections of the state, explaining how farmers would be benefitted by joining societies affiliated with the Bureau.[63] He was a persuasive orator, and his efforts were often successful. By June, 1858, as a result of his appeals new societies had been organized in nineteen counties, and at least one old association had been reactivated. In paying tribute to Williams, the editor of the Jackson *Mississippian* wrote in the summer of 1858:

We are satisfied, ere his labours are ended, there
will be an Agricultural Society in every county in
Mississippi, each numbering an average of 100 members, being a total of 6000 members in the State. When
these sub-societies become fully organized, and enter
into the spirit of their work, our State Agricultural
Bureau will be a most important feature of the Government, and Mississippi's wealth and resources will be
well known and felt.[64]

Williams's speeches to farm audiences over the state followed
the general pattern of a speech he delivered in Kemper County
in June, 1858. In this address he reminded cotton growers that
it was middlemen, not producers, who were establishing prices
for cotton. He urged them to organize so that they could wrest
control of the cotton market away from the middlemen and
consumers, and he offered the services of the Bureau to coordinate their efforts. In closing, Williams complimented his agrarian
audience by speaking of the "dignity of agriculture as a pur-
suit." It was, he told a cheering crowd, as important a science
as "Medecine [sic], Law, or Civil Engineering." [65]

The substructure of the Agricultural Bureau took shape
rapidly during the summer and fall of 1858, and by November
all was in readiness for its first state fair. This was held at
Jackson for four days beginning November 11. Attendance was
large; on a single day more than 5000 spectators visited its
various exhibits. Livestock shows and "trials of harness and
saddle horses" dominated the first three days' program. The
final day was devoted to a new feature: a ring tournament in
which gaily dressed riders demonstrated their horsemanship
and their skill with lances in competition for the prize of a
diamond ring. This event was received so enthusiastically by
the spectators that it was included in the program of all subsequent ante-bellum Mississippi State Fairs.[66]

Encouraged by this example and aided by funds from the
state treasury, local societies affiliated with the Bureau began
to hold county fairs at scattered points about the state.[67] These
were less elaborate than the state fair and they stressed home-

grown products and livestock rather than entertainment, but they were popular with country folk. Families often traveled to fairs in neighboring counties as well as to those given by their own societies, and it became customary to stagger schedules of fairs in order to avoid competition for audiences. Within three years, these county fairs won an important place in the social life of the agricultural population of Mississippi.[68]

The work of the Mississippi Agricultural Bureau was not limited to holding fairs and perfecting a state-wide farm organization. It was active also in a movement to create an association of planters covering all the cotton states.[69] With this purpose in mind, President Hudson in the summer of 1859 called on the planters of the Lower South to send delegates to a convention to be held at Nashville in October. In his appeal, he explained the objects of the gathering in these words:

> It is impossible in a communication like this to enumerate the advantages to be derived from annual meetings of intelligent representatives from the slave producing [sic] states. At these Conventions we can learn the efforts made annually to promote Agriculture and Mechanical interests of the different states, and the success of such efforts, determine what State legislation is necessary to advance these interests, the best manner of collecting and publishing Agricultural Statistics.[70]

The meeting convoked by the appeal of the Mississippi Agricultural Bureau met at Nashville in October, 1859, and, under the presidency of Hudson, became almost immediately a forum for discussion of secession.[71] And there is every reason to believe that this was in accordance with the wishes of Hudson and his Mississippi associates.[72] They had already attempted to use the state farm organization to promote secession in Mississippi. Williams, the permanent secretary, apparently had gone so far as to try to turn the local agricultural societies into covert secessionist cells. From the platform of his position within the Bureau, he frequently enunciated his views on the hopelessness of peaceful settlement of the South's controversy with the Northern states. His statement published in the Jackson *Mississippian* on the subject of the prospective Nashville convention left no doubt as to his frame of mind:

> For all the evils that effect [*sic*] the South, we have always been running to political conventions, and the clap trap of party machinery, when the true panacea lies in the *soil*, wanting only the skill to extract it. What boots it when we hurl our anathemas at the Black Republicans, when the pen, the paper and the table even . . . are made by them? What nonsense to be annunciating declarations of independence to all free-soildom when every implement of husbandry, every household and table comfort, with a few solitary exceptions, with all the show and equipage of pride, remind us of a despicable serfdom. We are not an indifferent spectator to the dangers that menace our national quiet, and feel as determined to throw the weight of our influence for the protection of our Constitutional, State, and *Southern* rights, should it, or submitting to the reign of Black Republicanism be an alternative; yet without the independence which arises from our own self-supplied wants and necessaries . . . [political independence is hopeless].[73]

Williams did not stop with championing economic independence. As secretary of the Bureau, he labored earnestly to promote the organization of local militia companies in conjunction with county agricultural societies, and in 1860 he went so far as to call for a state "encampment" of militia units as part of the Bureau's fair. To foster the "military spirit of our people," he suggested that the Bureau offer premiums of $300 and $150 for the best drilled units assembled at the fair. "Never," he wrote in 1860, "was the military spirit of our people so active, nor was there ever so heavy a tax laid upon the people to sustain it with the necessary equipment. . . ." There must be some test on the order of the proposed military encampment to see how far preparation for "the evil times ahead" has progressed, he continued, and "in no other way, and at no other time could it be better effected. . . ." [74]

A few shrewd newspaper editors who opposed secession were quick to sense the direction that the Agricultural Bureau was taking. The editor of the Kosciusko *Chronicle*, for example, immediately took exception to Williams's plan. "Now our idea of these Agricultural Fairs is that they are to promote agricultural and kindred objects," he wrote in some heat, "and we protest against their being turned into grand filibuster movements." [75]

Anti-secessionists were able to turn public opinion against the warlike program of the Agricultural Bureau, and Williams was compelled eventually to back away from his proposal. On May 1, 1860, the Democratic Jackson *Daily News*, stated in an editorial that Williams had made his suggestion as a private citizen and not in his official capacity as secretary of the state agricultural bureau. And the editor also strongly denied that Williams had intended that state funds be used to provide premiums for military units. [76] Public criticism of the proposed encampment continued to mount despite the efforts of Williams and his associates to quiet it, and the Bureau consequently was forced to abandon its military preparedness program altogether. [77]

This incident was embarrassing to the militant secessionists, because it exposed the manner in which the Agricultural Bureau had been subverted to partisan purposes. It is doubtful, however, that this revelation came in time to prevent this clique from setting up a state-wide secessionist organization under the guise of a harmless agricultural association. Evidence is lacking on this aspect of the 1860 secessionist movement, but it seems altogether possible that the local agricultural societies affiliated with the state Agricultural Bureau were the self-same vigilante societies and so-called "Committees of Public Safety" that did yeoman service in suppressing Unionist opposition to secession in 1860.

On the eve of the Civil War, cotton growers of Mississippi had good reason for optimism about their economic future. Their agricultural system based upon slave labor and cultivation of cotton had weathered every storm during its sixty-eight year existence, emerging from embargoes, wars and a great depression in a healthy and increasingly prosperous condition. Linked as it was with the British textile industry, it had shared in the irregular but phenomenal growth of that pioneering business phase of the Industrial Revolution.

To farmers and planters of the immediate pre-war generation, the depression of 1837-49 had been the acid test of their economic system, a test which slave-grown cotton had passed with flying colors. Indeed, the depression had been a blessing in disguise for the agricultural community of Mississippi, though not for the cotton growers of the older states to the east. Tobogganing cotton prices had been a powerful challenge, and Southwesterners had responded to it energetically and resourcefully. Faced with financial disaster, many of them had shaken off their characteristic conservatism and had learned to meet strange situations with freshly-devised expedients. Agriculturists who had been content to farm with the hoe as their fathers and grandfathers had done before them became experimenters and inventors of no mean ability during and after the depression.

Cotton growers, large and small, after 1839 began consciously to seek more effective ways of raising cotton and corn, discovering by experience how to make and use labor-saving farm implements and machinery. Having been made painfully aware of the need for thrift, Mississippians developed the habit of husbanding their financial resources by raising livestock and foodstuffs and by manufacturing on the plantation many of the articles they needed. Because clearing "new ground" had become uneconomical, planters were compelled during the depression to turn some of their attention to conserving the productivity of their cultivated fields. Erosion was checked by systems of scientifically planned drainage ditches and by rows plowed along contour lines. By trial and error, they developed practical and inexpensive systems of crop rotation, in which legumes were used to offset the drain that cotton and corn placed on the fertility of the soil. Furthermore, they greatly improved the quality if not the productivity of their cotton by conscientious application of the principles of selective breeding. Consequently, Southwestern cotton growers learned how to obtain greater proportional returns from their capital investments than they had been able to get even when cotton was selling for twenty cents a pound. This was no mean achievement!

Because of their improved cottons, implements and methods of cultivation, Mississippians in 1860 were able to produce more cotton with a given labor force—and produce it more cheaply—than in the pre-depression era. Furthermore, improved soil conservation techniques made it profitable to use their lands for much longer periods of time. Thus the road to wealth apparently was wide open to cotton growers in 1860, provided that the price of fiber remained in the ten to twelve cent range.

So far as Mississippians could see in 1860, there were only three threats to their continued prosperity, aside from the unlikely event of another depression. These were declining soil fertility, insects, and hostility of Northern abolitionists and anti-slavery politicians to their system of labor. To many the first two did not appear to be particularly dangerous. Insects, in those days before the arrival of boll weevils, were considered bothersome pests but not destroyers of crops. Diminishing productivity of the land was much more serious, in contemporary opinion, yet its progress was so slow that its consequences appeared to lie well in the future. Besides, crop rotation and newly-devised methods of soil conservation gave promise of eventually bringing this danger under control. Furthermore, even if the worst did come to pass, there still were hundreds of thousands of acres of fertile land suitable for cotton in the Yazoo-Mississippi Delta and across the Mississippi River in Louisiana, Arkansas and Texas. The real danger to their agricultural system, as a growing number of cotton producers saw it, was from the Northern abolitionists. This threat in the late 'fifties caused a growing number of planters and farmers to favor withdrawal from a Union they could no longer control as the best means of safeguarding their economic prosperity. In these years, cotton growers tended to shift their attention away from purely agricultural problems and to focus their thoughts upon politics—the area of greatest danger to their interests.

By 1860 perhaps a majority of Mississippians had become convinced that the Northern anti-slavery forces would soon succeed in imposing their will upon the nation as a whole. In their view the best means of protecting their economic prosperity was to withdraw from a union they no longer dominated. As this group was in control of the machinery of government, secession of the state followed as a matter of course. With this momentous decision the cotton growers of Mississippi staked their future—for better or worse—on political independence rather than continued progress in scientific agriculture.

Notes

FOOTNOTES TO CHAPTER ONE

1. Benjamin L. C. Wailes, *Report on the Agriculture and Geology of Mississippi, Embracing a Sketch of the Social and Natural History of the State* (Philadelphia, 1854), p. 128. Cited hereafter as Wailes, *Report*.

2. Clarence W. Alvord, *The Mississippi Valley in British Politics: A Study of the Trade, Land Speculation, and Experiments in Imperialism Culminating in the American Revolution* (Cleveland, 1917), II, 167–77; John F. H. Claiborne, *Mississippi, as a Province, Territory and State, with Biographical Notices of Eminent Citizens* (Jackson, Miss., 1880), p. 95. Cited hereafter as Claiborne, *Mississippi.*

3. Eron O. Rowland, ed., "Peter Chester," Mississippi Historical Society *Publications*, Centenary Series, V (1925), 77–78.

4. Eron O. Rowland, ed., *Life, Letters and Papers of William Dunbar of Elgin, Morayshire, Scotland, and Natchez, Mississippi: Pioneer Scientist of the Southern United States* (Jackson, Miss., 1930), p. 64. Cited hereafter as Rowland, *Dunbar.*

5. Arthur P. Whitaker, *The Spanish-American Frontier: 1783–1795: The Westward Movement and the Spanish Retreat in the Mississippi Valley* (Boston and New York, 1927), pp. 8–9. Cited hereafter as Whitaker, *The Spanish-American Frontier.* Claiborne, *Mississippi*, pp. 117–34.

6. This matter is discussed in great detail in Whitaker, *The Spanish-American Frontier.*

7. Wailes, *Report*, pp. 129–31: Rowland, *Dunbar*, pp. 23–70.

8. Whitaker, *The Spanish-American Frontier*, pp. 33–46, 103–107.

9. *Ibid.*, pp. 101–102

10. *Ibid.*

11. Arthur P. Whitaker, *The Mississippi Question, 1795–1803: A Study in Trade, Politics, and Diplomacy* (New York and London, 1934), pp. 62–3. Cited hereafter as Whitaker, *The Mississippi Question.*

12. Wailes, *Report*, pp. 132–34; Claiborne, *Mississippi*, pp. 135–36; Whitaker, *The Mississippi Question*, p. 61.

13. Whitaker, *The Spanish-American Frontier*, pp. 158–60.

14. *Ibid.* The text of a petition addressed to the Governor-General of New Orleans by many of the larger planters of Natchez relating to the history of their economic difficulties is published in Claiborne, *Mississippi*, pp. 139–40.

15. Whitaker, *The Spanish-American Frontier*, pp. 159–60

16. Claiborne, *Mississippi*, pp. 137–39.

17. Wailes, *Report*, pp. 135–39.

18. Claiborne, *Mississippi*, p. 140.

19. *Ibid.*

20. Information about roller gins given to B. L. C. Wailes by Christopher Miller, who settled in the Natchez region in 1783, is recorded in Wailes's unpublished multi-volume diary under the date line of February 25, 1853. These manuscripts are found in the library of Duke University and in the Mississippi Department of Archives and History. This historical source will be cited henceforth as Wailes, Diary.

21. Wailes, *Report*, pp. 155–57.

22. *Ibid.*, pp. 170–73; Jeannette Mirsky and Allan Nevins, *The World of Eli Whitney* (New York, 1952), p. 69.

23. Claiborne, *Mississippi*, p. 140 n.

24. Whitney's description of his gin, dated March 14, 1794, is published in Wailes, *Report*, pp. 367–69.

25. Wailes, *Report*, pp. 159–61.

26. Daniel Clark, Sr., to Colonel Anthony Hutchins, August 21, 1795. Quoted in part in Claiborne, *Mississippi*, p. 143

27. *Ibid.*; Wailes, *Report*, p. 167; Wailes, Diary, March 1, 1853; April 5, 1853; April 25, 1853.

28. Wailes, Diary, March 1, 1853; April 5, 1853; April 25, 1853; Claiborne, *Mississippi*, p. 143.

29. Wailes, *Report*, p. 167; Wailes, Diary, April 5, 1853.

30. John Hutchins, son of Anthony Hutchins, told Wailes that his father and William Voursdan visited the Clark gin while it was under construction. Wailes, Diary, April 5, 1853. Dunbar also recorded a visit made to that same gin in September, 1795. Wailes, *Report*, p. 167 n.

31. Wailes, Diary, March 1, 1853

32. *Ibid.*, March 25, 1853.

33. Claiborne, *Mississippi*, p. 143 n.; Wailes, *Report*, p. 168; Wailes, Diary, April 25, 1853.

34. Wailes, Diary, March 1, 4, and 12, 1853.

35. Claiborne, *Mississippi*, p. 143; Wailes, *Report*, pp. 168–69.

36. James Hall, *A Brief History of the Mississippi Territory, to which is Prefixed a Summary View of the Country between the Settlements on Cumberland River and the Territory* (Salisbury, N.C., 1801) reprinted in Mississippi Historical Society *Publications*, IX

(1906), 555. Cited hereafter as Hall, *A Brief History of the Missis-sippi Territory.*

37. Wailes, *Report,* pp. 168–69.

38. Wailes, Diary, March 1, 1853.

39. Hall, *A brief History of the Mississippi Territory,* p. 555; Wailes, Diary, March 1, 1853.

40. Francis Baily, *Journal of a Tour in Unsettled Parts of North America in 1796 & 1797* (London, 1856), p. 285. Cited hereafter as Baily, *Journal of a Tour.* Hall, *A Brief History of the Mississippi Territory,* p. 555.

41. Wailes, Diary, March 1, 1853; Wailes, *Report,* p. 169.

42. Wailes, Diary, March 1, 1853; April 5, 1853; Wailes, *Report,* p. 170.

43. Wailes, Diary, March 1, 1853.

44. Specimens of these negotiable cotton receipts are included in the Abijah and David Hunt Papers, Mississippi Department of Archives and History. This collection of manuscripts also contains information pertaining to the operation of a public ginning enter-prise during the early 1800s. Cited hereafter as Hunt Papers.

45. Wailes, Diary, March 1, 1853; Claiborne, *Mississippi,* p. 300; Harry Toulmin, *The Statutes of the Mississippi Territory, Revised and Digested by the Authority of the General Assembly, 1807* (Natchez, 1807), pp. 232–36.

46. The journal of an unidentified visitor to Natchez in 1799 was included in *Cuming's Tour to the Western Country,* 1807–1809, republished in Reuben Gold Thwaites, ed., *Early Western Travels, 1748–1846* (Cleveland, 1907), IV, 356–57.

47. William Dunbar to John Ross, May 23, 1799, published in part in Claiborne, *Mississippi,* p. 143.

48. N. Hunter, *Charges against Governor Sargent,* May 28, 1800, republished in Clarence E. Carter, ed., *The Territorial Papers of the United States* (Washington, 1937), V, 101. Cited hereafter as Carter, *Territorial Papers.* W. C. C. Claiborne to James Madison, December 20, 1801, in Dunbar Rowland, ed., *The Mississippi Territorial Archives, 1798–1803* (Nashville, 1905), I, 363–64. Hall, *A Brief History of the Mississippi Territory,* p. 555.

FOOTNOTES TO CHAPTER TWO

1. Sir George Watt, *The Wild and Cultivated Cotton Plants of the World: A Revision of the Genus Gossypium Framed Primarily with the Object of Aiding Planters and Investigators Who May Contemplate the Systematic Improvement of the Cotton Staple* (London, New York, Bombay and Calcutta, 1907), pp. 183–89. Cited hereafter as Watt, *Cotton Plants.*

2. *Ibid.*, pp. 183–84; "Historical and Statistical Collections of Louisiana—Parish of Catahoula," *De Bow's Review of the Southern and Western States*, XII (1852), 632–33.

3. R. B. Handy, "History and General Statistics of Cotton," *The Cotton Plant: Its History, Botany, Chemistry, Culture, Enemies and Uses* (Washington, D. C., 1896), p. 38; Wailes, *Report*, p. 156.

4. Lewis C. Gray, *History of Agriculture in the Southern United States to 1860* (Washington, D. C., 1933), II, 677.

5. Bienville and Salmon to Maurepas, May 12, 1733, published in Dunbar Rowland, ed., *Mississippi Provincial Archives 1704–1743* (Jackson, Miss., 1932), III, 600–601.

6. *Ibid.*

7. Claiborne, *Mississippi*, p. 140; "Historical and Statistical Collections of Louisiana," *De Bow's Review*, XII (1852), 632–33; Wailes *Report*, pp. 142–43; Watt, *Cotton Plants*, p. 120

8. Wailes, *Report*, pp. 142–43.

9. *Ibid.*; Watt, *Cotton Plants*, pp. 115–16, 265–95

10. "Historical and Statistical Sketches of Louisiana," *De Bow's Review*, XII, 632–33; Arthur Singleton, *Letters from the South and West* (Boston, 1824), pp. 113–14.

11. F. Cuming, *Sketches of a Tour to the Western Country Through the States of Ohio and Kentucky; A Voyage down the Ohio and Mississippi Rivers; and a Trip through the Mississippi Territory, and Part of West Florida, Commenced at Philadelphia in the Winter of 1807, and Concluded in 1809* (Pittsburg, 1810), pp. 281, 323; Hall, *A Brief History of the Mississippi Territory*, p. 555; Benjamin L. C. Wailes, "Cotton," *Southern Planter*, I (1842), 17–18.

12. Claiborne, *Mississippi*, p. 140; Abijah Hunt to Robert W. Gray, November 11, 1800, Hunt Papers; Wilkins & Linton to David Hunt, February 26, 1819, *ibid.*; E. J. Donnell, *Chronological and Statistical History of Cotton* (New York, 1872), pp. 71–109; *Mississippi Republican*, September 7, 1819.

13. Baily, *Journal of a Tour*, p. 285; "Historical and Statistical Collections of Louisiana," *De Bow's Review*, XII (1852), 632–33; Wailes, "Cotton," *Southern Planter*, I (1842), 17–18.

14. Thomas Affleck, "The Early Days of Cotton Growing in the Southwest," *De Bow's Review*, X (1851), 668–69.

15. William Dunbar to Mann & Barnard, September 14, 1804, Dunbar Papers, Mississippi Department of Archives and History; William Dunbar to Green & Wainwright, October 14, 1804, *ibid.*

16. Wailes, "Cotton," *Southern Planter*, I (1842), 17–18.

17. George F. Atkinson, "Diseases of Cotton," *The Cotton Plant*, p. 310; Wailes, *Report*, pp. 144–45.

18. Some of the numerous theories as to the cause of the rot

current during the ante-bellum period can be found in Wailes, "Cotton," *Southern Planter*, I (1842), 17–18; Wailes, *Report*, pp. 144–45; and in *Farmers Register*, I (1834), 575

19. Wailes, "Cotton," *Southern Planter*, I (1842), 17–18; Wailes, *Report*, pp. 144–45; *Farmers Register*, I (1834), 575; Singleton, *Letters from the South and West*, pp. 113–14.

20. "Historical and Statistical Collections of Louisiana," *De Bow's Review*, XII (1852), 632–33.

21. B. L. C. Wailes was the author of the only account now extant of the introduction of Mexican cotton into Mississippi. He obtained his information on this subject many years after the event in interviews with persons having first hand knowledge of the facts pertaining to this cotton. For Wailes's story of this event, see his *Report*, p. 143.

22. Dunbar described his new Mexican cotton as "of a fine rich color, very silky, fine & strong & rather a little longer–tho' not much–than our own staple. . . ." William Dunbar to Green & Wainwright, October 2, 1807, Dunbar Papers, Mississippi Department of Archives and History.

23. Samuel Postlethwait to Green & Wainwright, November 22, 1810, published in Rowland, *Dunbar*, pp. 389–90.

24. *American Farmer*, II (1820), 116.

25. Wailes, "Cotton," *Southern Planter*, I (1842), 17–18.

26. *Ibid.*

27. Affleck, "Early Days of Cotton Growing," *De Bow's Review*, X (1851), 668–69; *American Farmer*, II (1820), 116; Claiborne, *Mississippi*, pp. 140–42; Wailes, "Cotton," *Southern Planter*, I (1842), 17–18; Wailes, *Report*, pp. 143–44; Watt, *Cotton Plants*, pp. 21, 192–96.

28. Affleck, "Early Days of Cotton Growing," *De Bow's Review*, X (1851), 668–69.

29. Claiborne, *Mississippi*, pp. 140–42; H. S. Fulkerson, *Random Recollections of Early Days in Mississippi* (Vicksburg, Miss., 1885), pp. 12–14.

30. *Farmers Register*, II (1835), 122–23; James L. Watkins, *King Cotton: A Historical and Statistical Review, 1790–1908* (New York, 1908), pp. 74–75; Watt, *Cotton Plants*, pp. 226–39.

FOOTNOTES TO CHAPTER THREE

1. Hall, *A Brief History of the Mississippi Territory*, p. 555.

2. [Joseph H. Ingraham], *The South-West by a Yankee* (New York, 1835), II, 86–88.

3. "Preservation of the Soil," *Southern Planter*, I (1842), 13–14.

4. *Southern Cultivator*, XIII (1855), 219.

5.   Colonel Henry W. Vick to Samuel E. Bailey, October 4, 1842, published in *Southern Planter*, I (1842), 17–18.

6.   John H. Moore, ed., "Two Documents Relating to Plantation Overseers of the Vicksburg Region, 1831–1832," *Journal of Mississippi History*, XVI (1954), 31–36.

7.   *Southwestern Farmer*, I (1842), 100, 205.

8.   [Ingraham], *The South-West by a Yankee*, II, 101.

9.   The residence of John T. Leigh, owner of a medium sized cotton plantation in Yalobusha County, which Solon Robinson described in 1845 as "a common log cabin, with a hall in between," was typical of the plantation houses of Mississippi at that period. Herbert A. Keller, ed., *Solon Robinson: Pioneer and Agriculturist* (2 vols., Indianapolis, 1936), II, 458.

10.   James R. Creecy, *Scenes in the South, and Other Miscellaneous Pieces* (Washington, 1860), pp. 83–84.

11.   A typewritten copy of the unpublished diary of Edward R. Wells is in the possession of the Mississippi Department of Archives and History. Cited henceforth as Wells, Diary.

12.   Thomas Affleck to James D. B. De Bow, *De Bow's Review*, V (1848), 82–84.

13.   An early expression of the general Southwestern sentiment about the value of land is found in a memorial addressed to Congress in 1814 by the Territorial legislature of Mississippi protesting the use of landholding as a basis for voting qualifications. "That peculiar and settled interest which is presumed when the right of Election is confined to Landholders does not really exist. The man who has a large Stock of cattle, or a considerable Property in Slaves, has surely more interest in the country in which they are located than the man who owns fifty acres of Land which may not be worth as many cents, and are verry [sic] probably not worth more than One Hundred Dollars." Carter, *Territorial Papers*. VI, 411–13.

14.   An excellent contemporary analysis of the reasons underlying the Southern cotton grower's disregard for the long term value of the soil was written by C. H. Howard, an associate editor of the *Southern Cultivator*. Published in the United States Patent Office *Report: Agriculture* (1860), pp. 224–39.

15.   An excellent detailed description of farming methods then in use on Mississippi cotton plantations was written in 1835 by the historian and cotton planter John W. Monette and published in full in [Ingraham], *The South-West by a Yankee*, pp. 281–291. Cited hereafter as Monette, "The Cotton Crop," *The South-West by a Yankee*.

16.   "Historical and Statistical Collections of Louisiana," *De Bow's Review*, XII (1852), 632–33; Wailes, "Cotton," *Southern Planter*, I (1842) 17–18.

17. Monette, "The Cotton Crop," *The South-West by a Yankee,* II, 282–83.

18. George Dougharty to Benjamin L. C. Wailes, June 3, 1853, Wailes Papers, Duke University Library; Edward Turner to Benjamin L. C. Wailes, April 14, 1859, Wailes Papers, Mississippi Department of Archives and History; Wailes *Report,* p. 153.

19. Rowland, *Dunbar.*

20. *American Cotton Planter,* IV (1856), 347–48; *Southern Planter, Planter,* I (1842), 10; *Southwestern Farmer,* I (1842), 17, 73.

21. Monette, "The Cotton Crop," *The South-West by a Yankee,* II, 283–85.

22. Affleck, "Early Days of Cotton Growing," *De Bow's Review,* X (1851), 668–69; Cuming, *Sketches of a Tour,* pp. 281, 323.

23. Wailes, *Report,* pp. 142–43.

24. Carter, *Territorial Papers,* V, 101; "Historical Collections of Louisiana," *De Bow's Review,* XII (1852), 632–33.

25. In an attempt to set a cotton picking record, eleven Negroes on the plantation of A. W. Hutchins of Adams County gathered 3,316 pounds of seed cotton in one day. *Mississippi Free Trader and Natchez Gazette,* October 13, 1836.

26. In the years 1833 and 1836 Richard Nutt, son of Dr. Rush Nutt, averaged nine bales of cotton to the hand, a record almost equaled by his brother. *Southwestern Farmer,* I (1842), 116.

27. The following advertisement which appeared in the *Natchez,* January 22, 1831 was typical of the period: "Received on commission . . . Eighty small Ploughs of M'Cormack's Patent, made by Messrs. J. & L. Jacobs of Maysville [Kentucky]. Ferguson & Gillett."

28. On January 8, 1831, a Natchez plough manufacturer advertised three hundred of his implements for sale at his shop. The *Natchez,* January 22, 1831. In the newly settled northern parts of the state, however, cotton growers in 1836 were using the so-called "half share or twisting shovels" as the only turning plow, and this implement was a product of local blacksmith shops. J. M. Townes, "Plows and Ploughing," *Southern Cultivator,* V. (1847), 123.

29. William Dunbar to Green & Wainewright, April 2, 1806, published in Rowland, *Dunbar,* p. 337.

30. Wailes, *Report,* pp. 170–73.

31. Wailes, Diary, March 1, 1853.

32. Natchez *Mississippi Republican,* August 15, 1820.

33. Gray, *Agriculture in the Southern United States to 1860,* I, 542.

34. Monette, "The Cotton Crop," *The South-West by a Yankee,* II, 288–90.

35. *Ibid.*

36. William Dunbar to [?], October 20, 1805, Rowland, *Dunbar,*

p. 321; Claiborne, *Mississippi*, p. 141; Monette, "The Cotton Crop," *The South-West by a Yankee*, II, pp. 288–90.

37.   Claiborne, *Mississippi*, p. 141

38.   Monette, "The Cotton Crop," *The South-West by a Yankee*, II, 288–90.

39.   Wailes, *Report*, p. 173.

40.   *Ibid.*, p. 174.

41.   Claiborne, *Mississippi*, p. 144

42.   William Dunbar to Charles Ross, [1803], Dunbar Papers, Mississippi Department of Archives and History

43.   *Ibid.*

44.   Claiborne, *Mississippi*, p. 144.

45.   Monette, "The Cotton Crop," *The South-West by a Yankee*, 288–90.

46.   "The Cotton Press," *Southwestern Farmer*, I (1842), 77–78; Wailes, *Report*, p. 174.

47.   "The Cotton Press," *Southwestern Farmer*, I (1842), pp. 77–78; Wailes, *Report*, pp. 175–76.

48.   "Weight of Cotton Bales," Grand Gulf *Advertiser*, July 28, 1836; Wailes, *Report*, p. 177.

49.   [Ingraham], *The South-West by a Yankee*, II, 170–71.

50.   Abijah Hunt to Robert W. Gray, November 11, 1800, Hunt Papers, Mississippi Department of Archives and History.

51.   Edward Quick and Herbert Quick, *Mississippi Steamboating: A History of Steamboating on the Mississippi and its Tributaries* (New York, 1926), pp. 168–78.

52.   For additional information about transportation on the rivers of the state refer to the following: R. Baird, *View of the Valley of the Mississippi, or the Emigrant's and Traveller's Guide to the West* (Philadelphia, 1834), pp. 265–67; H. S. Fulkerson, *Random Recollections*, pp. 19–20; [Ingraham], *The South-West by a Yankee*, II, 174; Panola *Lynx*, January 11, 1845, November 29, 1845; Port Gibson *Herald*, December 28, 1843; Jackson *Southern Weekly Reformer*, December 26, 1843, November 11, 1844, January 3, 1845; Yazoo *Democrat*, October 8, 1845; Vicksburg *Weekly Sentinel*, November 24, 1847.

53.   Gray, *Agriculture in the Southern United States to 1860*, II, 706; Watkins, *King Cotton*, p. 194.

54.   [Ingraham], *The South-West by a Yankee*, II, 160, 169; Grand Gulf *Advertiser*, February 8, 1839

55.   Cumings, *Sketches of a Tour*, p. 323; Henry Ker, *Travels Through the Western Interior of the United States from the year 1808 up to the Year 1816* (Elizabethtown, N. J., 1816), p. 42; [Ingraham], *The South-West by a Yankee*, II, 89; Raymond *Times*, May 7, 1841; Christian Schultz, *Travels on an Inland Voyage through the States of New York, Pennsylvania, Virginia, Ohio, Kentucky and*

Tennessee, and through the Territories of Indiana, Louisiana, Missis-sippi and New Orleans, Performed in the Years 1807 and 1808 (2 vols., New York, 1810), II, 139.

56.   E. Mackenzie, *An Historical, Topographical, and Descriptive View of the United States of America and of Lower Canada* (New-castle upon Tyne, 1819), p. 290.

57.   Baird, *View of the Valley of the Mississippi,* pp. 264–65; Mackenzie, *An Historical, Topographical, and Descriptive View of the United States,* p. 290; Wailes, *Report,* pp. 181–97.

58.   Wailes, *Report,* pp. 181–86; Raymond *Times,* April 24, 1840.

59.   The method of cultivating corn used on the Killona Plantation was explained in great detail in the section devoted to the year 1838. Killona Plantation Journal, Mississippi Department of Archives and History. See also *Southern Cultivator,* V (1847), 124–25.

60.   The problems involved in the pulling of fodder from green corn were discussed at length by Thomas Affleck in a communication to the U. S. Patent Office. U. S. Patent Office *Report* (1849–50), pp. 153–54. Refer also to Wailes, *Report,* p. 184

61.   Raymond *Times,* May 7, 1841.

62.   Martin W. Philips, "Southern Crops and Culture," New Orleans *Commercial Times,* June 20, 1846.

63.   Claiborne, *Mississippi,* p. 141

64.   U.S. Patent Office *Report* (1849–50), pp. 152–67; *Southern Cultivator,* VI (1848), 150; *Southwestern Farmer,* I (1842), 58, 109, 116, 132

65.   The important place sweet potatoes occupied in Mississippi agriculture is demonstrated clearly by the following account: "Col. H. D. Robertson, near Clinton, Hinds Co., with ten hands, made and gathered last year 100 heavy bales of cotton; 3000 bushels of corn, and 1500 bushels of potatoes. He killed some seventy-five large hogs; stall fed three fine beeves. . . ." *Mississippi Free Trader,* April 19, 1843.

66.   Philips, "Southern Crops and Culture," New Orleans *Com-mercial Times,* June 20, 1846; Wailes, *Report,* pp. 190–93.

67.   Cuming, *Sketches of a Tour,* p. 323; Mackenzie, *An Historical, Topographical, and Descriptive View of the United States and of Lower Canada,* p. 290; Julia Ideson and Sanford W. Higginbotham, eds., "A Trading Trip to Natchez and New Orleans, 1822: Diary of Thomas S. Teas," *Journal of Southern History,* VII (1941), 387; Natchez *Mississippi Republican,* September 23, 1819.

68.   Information on the South Mississippi cattle country is found in the following sources: John F. H. Claiborne, "A Trip Through the Piney Woods," Mississippi Historical Society *Publications,* IX (1906), 487–538; *De Bow's Review,* XVII (1854), 627–28; Timothy Flint, *Recollections of the Last Ten Years* (Boston, 1826), pp. 317, 329; Grenada *Harry of the West,* November 23, 1844; Canton

AGRICULTURE IN ANTE-BELLUM MISSISSIPPI

Independent Democrat, December 2, 1843; Natchez *Mississippi Free Trader,* May 29, 1844; U.S. Patent Office *Report* (1849–50), pp. 152–67; *ibid.: Agriculture* (1850–51), pp. 257–61; Raymond *Times,* December 17, 1841; *Southwestern Farmer,* I (1842), 3; Wailes, Diary, August 10–15, 1852.

69. Eugene W. Hilgard, *Report on the Geology and Agriculture of the State of Mississippi* (Jackson, Miss., 1860), p. 361.

70. Mark R. Cockrill, "Sheep Raising in the South," *De Bow's Review,* XII (1852), 584–95; "Blooded Stock," *Southwestern Farmer,* I (1842), 118.

71. *Southern Planter,* I (1842), 18–19; Charles S. Sydnor, *A Gentleman of the Old Natchez Region: Benjamin L. C. Wailes* (Durham, 1938), pp. 82–83. Cited hereafter as Sydnor, *Benjamin L. C. Wailes.*

72. A very full description of hog raising on Mississippi plantations during the 'twenties and 'thirties appeared in 1839 in the short-lived Raymond *Mississippi Farmer,* the first agricultural periodical published in the Old Southwest. No copies of this journal are known to exist today, but some of the articles originally written for the *Farmer* were republished at a later date in other papers belonging to the same proprietor. *Southwestern Farmer,* I (1842), 124.

73. Raising of horses in Mississippi was described by Thomas Affleck in a communication to the U. S. Patent Office, dated November, 1849. U. S. Patent Office *Report* (1849–50), pp. 160–61.

74. [Ingraham], *The South-West by a Yankee,* II, 94; 172.

75. Slavery on plantations in Mississippi has been described and analyzed by Charles S. Sydnor, *Slavery in Mississippi* (New York and London, 1933).

76. Further details on the life and duties of overseers in Mississippi can be found in John S. Bassett, *The Southern Plantation Overseer as Revealed in His Letters* (Northhampton, Mass., 1925) and Moore, "Two Documents Relating to Plantation Overseers of the Vicksburg Region, 1831–32," *Journal of Mississippi History,* XVI (1954), 31–36.

77. Wailes, Diary, March 19, 1853.

78. [Ingraham], *The South-West by a Yankee,* II, 94.

79. *Ibid.,* p. 93.

80. Accounts of conditions in the "New Purchase" during the 1834–37 period have been published in Bassett, *The Plantation Overseer;* Joseph G. Baldwin, *The Flush Times of Alabama and Mississippi* (San Francisco, 1876); Creecy, *Scenes in the South;* and Fulkerson, *Random Recollections.*

81. The daily cotton prices prevailing in the New Orleans market for the period between 1825 and the Civil War were republished from New Orleans *Price Current* in Donnell, *History of Cotton.*

FOOTNOTES TO CHAPTER FOUR

1. Ulrich B. Phillips, *Life and Labor in the Old South* (Boston, 1948), p. 177.

2. Census Bureau, *Aggregate Amount of Persons within the United States in the Year 1810* (Washington, D. C., 1811), p. 83; *Abstract of the Returns of the Fifth Census, Showing the Number of Free People, the Number of Slaves, the Federal or Representative Number and the Aggregate of Each County of Each State of the United States* (Washington, D. C., 1832), p. 36.

3. Donnell, *History of Cotton*, pp. 104–233; Phillips, *Life and Labor in the Old South*, p. 177

4. Baldwin, *Flush Times of Alabama and Mississippi*, pp. 236–37.

5. The official returns of a census conducted by the state government of Mississippi in 1836 were published in various newspapers of the state. The statement enumerated the slaves and white persons, bales of cotton produced, and the acreage in cultivation for each county. *Mississippi Free Trader and Natchez Gazette*, May 18, 1837.

6. Donnell, *History of Cotton*, pp. 220–34; Samuel E. Morison and Henry S. Commager, *The Growth of the American Republic* (New York and London, 1942), I, 562–63; M. B. Hammond, *The Cotton Industry: An Essay in American Economic History* (New York and London, 1897), pp. 71–72.

7. *Compendium of the Enumerations of the Inhabitants and Statistics of the United States, as Obtained at the Department of State, from the Returns of the Sixth Census* . . . (Washington, D. C., 1841), pp. 58–59

8. Mississippi's agricultural expansion is reflected in the annual reports of land sold by the federal government during the 1833–45 period. The following statistics appeared in *De Bow's Review*, IV (1847), 86:

1833	1,221,494 acres
1834	1,064,054 "
1835	2,931,181 "
1836	2,023,709 "
1837	256,354 "
1838	271,074 "
1839	17,784 "
1840	19,174 "
1841	21,635 "
1842	43,966 "
1843	34,500 "
1844	34,436 "
1845	28,232 "

9. "This country was just settling up. Marvelous accounts had gone forth of the fertility of its virgin lands; and the productions

of the soil were commanding a price remunerating to slave labor as it had never been remunerated before. Emigrants came flocking in from all quarters of the Union, especially from the slaveholding States. . . . Under this stimulating process prices rose like smoke. Lots in obscure villages were held at city prices; lands bought at the minimum cost of government, were sold at from thirty to forty dollars per acre, and considered dirt cheap at that. In short, the country had got to be a full ante-type of California; in all except the gold." Baldwin, *Flush Times of Alabama and Mississippi*, pp. 82–84.

10.   "Speculation, speculation, has been making poor men rich and rich men princes; men of no capital, in three years time have become wealthy and those of some have grown to hundreds of thousands." William H. Wills, "A Southern Traveler's Diary, 1840," Southern History Association *Publications*, VIII (1904), 36–37.

11.   *Ibid.;* Baldwin, *Flush Times of Alabama and Mississippi*, pp. 90–91.

12.   Donnell, *History of Cotton*, pp. 162–64.

13.   Baldwin, *Flush Times of Alabama and Mississippi*, p. 240; Bassett, The *Southern Plantation Overseer*, p. 136; J. A. Orr, "A Trip from Houston to Jackson, Miss., in 1845," Mississippi Historical Society *Publications*, IX (1906), 175; Franklin L. Riley, ed., "Diary of a Mississippi Planter," Mississippi Historical Society *Publications*, X (1909), 318; Willis, "A Southern Traveler's Diary, 1840," Southern History Association *Publications*, VIII (1904), 32.

14.   Orr, "A Trip from Houston to Jackson, Miss., in 1845," Mississippi Historical Society *Publications*, IX (1906), 175.

15.   Baldwin, *Flush Times of Alabama and Mississippi*, pp. 90–93; Orr, "A Trip from Houston to Jackson, Miss., in 1845," Mississippi Historical Society *Publications*, IX (1906), 176; Riley, ed., "Diary of a Mississippi Planter," Mississippi Historical Society *Publications*, X (1909), 318.

16.   In an address delivered to the members of the Agricultural, Horticultural and Botanical Society of Jefferson College at Washington, Mississippi, on April 28, 1843, President B. L. C. Wailes remarked: "Happily, we have learned wisdom enough in our adversity, to provide at home many of the necessaries of life; but the extent to which we are still dependent on foreign supply, in the circumstance of the country, is appalling. The cost of horses and mules, farming implements, bagging and rope, pork and clothing, still requires a most formidable aggregate drained from our resources." *Southern Cultivator*, I (1843), 92–93.

17.   Jackson *Southron*, August 26 and September 9, 1841

18.   The history of one unsuccessful effort to launch an agricultural periodical can be gleaned from a number of letters and articles appearing in the Jackson *Southron*, December 23, 1841, January 20, 1842, March 5, 1842, April 28, 1842, May 2, 1842.

19. While Solon Robinson was traveling through Mississippi in 1845 as a roving correspondent for the Albany (New York) *Cultivator*, he visited N. G. North at the offices of the *Southwestern Farmer* at Raymond, in company with Martin W. Philips. On this occasion, North told Robinson that his list of subscribers, after years of publication, had reached "almost to four hundred." Kellar, *Solon Robinson*, II, 479–80

20. North purchased a political newspaper, the Raymond *Times*, from Samual T. King in 1839 and continued to publish it under that title until the spring of 1842. While still publishing the *Times*, North put out three issues of an agricultural periodical, called the *Mississippi Farmer*, with the object of determining whether the planters of the state would be willing to support such a journal. At that time North obtained too few promises of subscriptions to justify the experiment, but he did win the backing of a number of the leaders of the nascent agricultural reform movement. Finally, in 1842, becoming convinced that an agricultural paper could be made to pay, North converted the political Raymond *Times* into the nonpolitical agricultural journal, the *Southwestern Farmer*.

21. Unfortunately none of the three issues of the state's pioneer agricultural paper, the *Mississippi Farmer*, has been preserved. A few of its better articles were reprinted in the *Southwestern Farmer* at a later date, and the following table of contents for these lost issues was printed as an advertisement in the Raymond *Times*, January 3, 1840:

"Contents of No. 1.—Dec. 2, 1839

Prospectus. Introductory Address. Communications: Cultivation of Millet, by R. A. Patrick; Food for Hogs, by an Enquirer. Agricultural Reform in South Carolina. Map and Directory of Hinds County. The Silk Culture—Extracts from *Cheraw Gazette* and Other Publications on that Subject. Proverbs Pertaining to Agriculture. Southern Gardener's Calendar. Petitions in Behalf of the Agricultural Interest to the Legislature of Tennessee and Pennsylvania. Editorial and Commercial Articles. Death of Judge Buel. Miscellany.

Contents of No. II.—Dec. 16, 1839

Profits of Science for Farmers. Communications: The Art of Curing Bacon, by Col. T. S. Dabney; Food for hogs, by John Jenkins; Clover, by a Young Farmer. History of the Cotton Culture. Extracts Concerning Ruta Baga, Mangel Wurtzel, the Sugar Beet, The Comparative Value of Hay, Vegetables, and Corn, The Goose Wheat, Pork and Bread Stuffs at the West. Millet. Butter. Restoring of Peach Trees. Soil of the Sea Coast Counties of Mississippi. &c. &c. Notice of Dr. New's Address. Editorials. Miscellany.

Contents of No. III.—Jan. 6, 1840.

Legislative Aid to Agriculture. Projected State Agricultural Convention. Communications: Lucerne, by M. W. Philips; On the Manufacturing of Our Shoes, &c., &c., by J. H. Johnson; On Raising Hogs, by 'H', On Preserving Sweet Potatoes, by S. C. Barton; friendly letters from John Gilmer, Esq., and 'Farmer Giles.' Sheep in Spain. Blooded Stock for Hinds County. Arrivals of Berkshire Pigs, &c., &c., at New Orleans. Miscellany."

22. John C. Jenkins, Jr., enjoyed a very successful career as a Mississippi journalist until his death at the hands of an assassin near the close of the depression decade. He served for a time as associate editor and engraver for the *Southwestern Farmer,* and then became editor and proprietor of the Vicksburg *Sentinel.* As son of John C. Jenkins, Sr., a well-known Southern writer on horticulture, young Jenkins was always deeply concerned with problems of Mississippi agriculture, and he frequently wrote upon agricultural subjects in both political and agricultural papers. He, Philips, and North were active in the organization of agricultural societies in Hinds and Madison counties, and in 1842 they were the moving spirits behind the campaign for a state wide society.

23. Martin W. Philips, of Edwards, Mississippi, was the most prolific American agricultural writer of the ante-bellum period. Philips was an indefatigable if rather unscientific experimenter in all matters pertaining to the agriculture of the lower South, and he made a habit of communicating his views and the results of his experiments to a vast number of political and agricultural journals in all sections of the United States. All these letters, running into the hundreds, are valuable repositories of information about current agricultural practices of the cotton growing states. A brief sketch of Philips's career was published along with a portion of his diary in Franklin L. Riley, ed., "Diary of a Mississippi Planter," Mississippi Historical Society *Publications,* X (1909), 305–309.

24. The only extant volume of the *Southwestern Farmer* is in the Mississippi Department of Archives and History.

25. The prospectus of the *Southwestern Farmer,* dated February 7, 1842, was published in the Raymond *Times,* February 25, 1842, and a second prospectus, dated November 1, 1844, announcing a change in format, was published in the Vicksburg *Sentinel,* February 10, 1845.

26. An obituary for the recently deceased *Southwestern Farmer* appeared in Jenkins's Vicksburg *Sentinel,* August 8, 1845.

27. B. L. C. Wailes's bound copy of all the issues of the *Southern Planter* is preserved in the Mississippi Department of Archives and History.

28. A short account of Bailey's career in Natchez is written on the fly leaves of the bound volume of the *Southern Planter* belonging to the Mississippi Department of Archives and History. The hand-writing appears to be that of B. L. C. Wailes

29. The names of new subscribers were listed in each issue of the *Southern Planter*. The total number was 226.

30. Vicksburg *Register,* September 10, 1845; Woodville *Republican,* October 25, 1845.

31. *Cultivator,* IV (1837), 1.

32. *Mississippi Free Trader and Natchez Gazette,* April 29, 1837.

33. *Ibid.,* March 31, 1842; *Southwestern Farmer,* I (1842), 37.

34. Jackson *Southron,* March 23, 1843.

35. In 1842 the *Southwestern Farmer* quoted from fifteen out-of-state agricultural periodicals. At the same time the *Southern Planter* was receiving a total of twenty-two.

36. Almost complete files of the *Southern Cultivator* are to be found in the library of the University of Georgia and in the Carnegie Library, Atlanta, Georgia. For a brief history and description of this paper, refer to Albert L. Demaree, *The American Agricultural Press 1819–1860* (New York, 1941), pp. 372–75.

37. The first serious effort to organize an agricultural society in Mississippi had met with failure. On December 29, 1838, many farmers of Marshall County, in the northern part of the state, had convened in the town of Holly Springs with the object of forming an agricultural organization, but for some unknown reason the meeting had ended without concrete results. In reporting the incident, a local newsman had expressed regret at the failure of the gathering, and stated his opinion on the value of such organizations as follows: "We are well aware that Agricultural Societies have been a blessing to the whole farming community. They have improved the great field of agriculture, have distributed among the yeomen of the land a variety and amount of information upon agricultural subjects; they have dispensed great light to farmers generally—as to the best breeds to be selected in the animal kingdom—the best manner of raising the different livestock; the best economy of feeding, etc.; they have also been the means of turning to the general advantage the results of successful experiments made by scientific agriculturists in the vegetable kingdom. Holly Springs *Marshall County Republican and Free Trade Advocate,* January 12, 1839.

38. For discussions of the Southern agricultural reform movement during the ante-bellum period, see James C. Bonner, "Advancing Trends in Southern Agriculture, 1840–1860," *Agricultural History,* XXII (1948), 248–59; and "Genesis of Agricultural Reform in the Cotton Belt," *Journal of Southern History,* IX (1943), 475–500; see also Avery O. Craven, "The Agricultural Reformers of the Ante-Bellum South," *American Historical Review,* XXXIII (1927–28), 302–14.

39. Natchez *Mississippi Republican,* September 7 and 23, 1819; November 2, 1819.

40. Information on the ante-bellum history and cultural activities

of Jefferson College can be found in Sydnor, *Benjamin L. C. Wailes*, pp. 204–33.

41. The proceedings of the organizational meeting of the Jefferson College agricultural society were published in the *Mississippi Free Trader and Natchez Weekly Gazette*, May 23, 1839.

42. *Southern Planter*, I (1842), 78.

43. The basic attitude of the membership of this society was revealed in two addresses made by its president B. L. C. Wailes that were published in pamphlet form at the time. Excerpts from both appeared in several newspapers and agricultural journals. Benjamin L. C. Wailes, *Address Delivered in the College Chapel before the Agricultural, Horticultural, and Botanical Society, of Jefferson College on the 29th of April, 1841* (Natchez, 1841), and *Address Delivered at Washington, Miss., before the Agricultural, Horticultural and Botanical Society of Jefferson College on the 29th of April, 1842* (Natchez, 1842).

44. Sydnor, *Benjamin L. C. Wailes*, pp. 158–59

45. Descriptions of fairs held by the Jefferson College agricultural society were published in the following: *Mississippi Free Trader and Natchez Gazette*, August 8, 1839, November 29, 1841, May 5, 1842, November 23, 1842, April 28, 1843, May 3, 1843; *American Agriculturist*, IV (1845), 17–18; Raymond *Times*, August 16, 1839; *Southwestern Farmer*, I (1842), 105–106.

46. *Mississippi Free Trader and Natchez Gazette*, May 5, 1842; *Southwestern Farmer*, I (1842), 105–106.

47. *Mississippi Free Trader and Natchez Gazette*, May 5, 1842; *Southern Planter*, I (1842), 12–13. For an account of the first hogs imported into the Natchez area see the Raymond *Times*, September 24, 1841.

48. *Mississippi Free Trader and Natchez Gazette*, May 5, 1842; *Southwestern Farmer*, I (1842), 121.

49. *American Agriculturist*, IV (1845), 17–18.

50. *Ibid.*

51. The lists of members of the Jefferson College agricultural society given in the Natchez press in 1839 reveal that almost all of them were of the large planter class, and in the Natchez region the custom was for overseers to direct farming operations on plantations that had sizeable forces of Negro slaves. Thomas Affleck, however, was a marked exception to the rule. In the early 1840s, he was occupied in building up an orchard and nursery on a very small plantation containing only forty acres, situated near the village of Washington. Furthermore, Affleck differed from his fellow members of the society in the respect that he had received an education in horticulture under the best authorities in Scotland, and he was interested in applying scientific methods to the improvement of agricultural and horticultural practices.

52. Raymond *Times,* June 7, 1839.

53. *Ibid.,* June 6, 1839.

54. *Ibid.*

55. *Ibid.,* June 22 and 28, 1839, September 17 and 20, 1839, December 20 and 27, 1839.

56. The proceedings of agricultural meetings at Clinton and Raymond were published in the Raymond *Times,* June 28, 1839.

57. Raymond *Times,* August 2, 1839.

58. Among others, county agricultural associations were formed at this time at Canton (Madison County), Macon (Noxubee County), Aberdeen (Monroe County), and at Paulding (Jasper County). For detailed information on these organizations, see Aberdeen *Whig* and *North Mississippi Advocate,* June 11, 1839; Macon *Intelligencer,* August 29, 1839, September 5, 1839, October 3 and 10, 1839; *Mississippi Free Trader and Natchez Gazette,* June 20, 1839, September 17, 1839, and May 11, 1840.

59. Raymond *Times,* December 20 and 27, 1839.

60. *Ibid.,* December 27, 1839.

61. *Ibid.,* January 24, 1840.

62. A number of persons at the Jackson convention, for example, promised to subscribe to North's *Mississippi Farmer,* but none of them was sufficiently interested in such matters to mail his advance payments to the office of the Raymond *Times* in accordance with his promise. Raymond *Times,* January 1, 1841. For further information on the early history of the Mississippi State Agricultural Society, refer also to the *Southwestern Farmer,* I (1842), 73.

63. The experience of the Monroe County association, described in the following words by John L. Tindall, was typical of all similar organizations of the period, with the sole exception of the Jefferson College society: "More than two years since, a few of our citizens organized at this place 'The Monroe County Agricultural Society,' but few have taken any interest in it, and consequently its doings have been very limited, and the detail would be uninteresting; its friends are looking forward to a brighter day as we now have in this vicinity a number of sensible and energetic planters. . . ." *Southern Planter,* I (1842), 11–12.

64. Martin W. Philips to the editor of the Canton *Mississippi Creole,* reprinted in the Raymond *Times,* July 16, 1841.

65. *Ibid;* Jackson *Southron,* December 23, 1841.

66. In this period local agricultural societies were organized in Jefferson, Lowndes, Warren, Wilkinson, Jasper and Yalobusha Counties. For detailed information see Grenada *Register,* June 4, 1842, July 4, 1842, October 15, 1842; Jackson *Southern Reformer,* May 11, 1844, May 8, 1845, July 15, 1845; Jackson *Southron,* August 26, 1841; *Mississippi Free Trader and Natchez Gazette,* May 24, 1843, May 15, 1844; Paulding *True Democrat,* July 30, 1845; Ripley *Advertiser,*

May 25, 1844; *Southwestern Farmer,* I (1842), 22–155; Vicksburg *Sentinel,* May 6, 1844, July 30, 1845.

67.   At the first meeting held on January 21, 1842, the members of the state agricultural society voted in favor of holding a state fair in the following April and elected the following officers: Pryor Lea of Hinds County as president; Colin S. Tarpley of Hinds County as corresponding secretary; Dr. Joseph S. Copes of Hinds County as recording secretary; and James Elliot as treasurer. James Rucks, Hiram C. Runnels, W. M. Rives, Daniel Mayes, John B. Peyton, Benjamin Ricks, and Dr. Farrar made up the executive committee. T. M. Tucker of Lowndes, W. L. Sharkey of Warren, Martin W. Philips of Hinds, Edward Turner of Franklin, Arthur Fox of Lawrence, J. J. Moore of Noxubee, James F. Trotter of Marshall, John W. Kendall of Carroll, Eli T. Montgomery of Madison, and B. L. C. Wailes of Adams, were elected vice presidents. *Southwestern Farmer,* I (1843), 1–13.

68.   As a typical example, the Jefferson County Agricultural Society, organized at Fayette on March 24, 1842, held local fairs in that town in 1842, 1843, 1844 and 1845. Competition between various classifications of jacks, horses, mules, cows, sheep and hogs were the main events at these fairs, and they usually drew respectable crowds. In the fall of 1842, for example, the fair was visited by more than a thousand spectators. Greenwood *Reporter,* April 22, 1845; *Mississippi Free Trader and Natchez Gazette,* March 14, 1842, May 19, 1843; Port Gibson *Herald,* May 18, 1843; *Southwestern Farmer,* I (1842), 116–17.

69.   Jackson *Southron,* May 31, 1843; *Mississippi Free Trader and Natchez Gazette,* June 7 and 14, 1843; Washington *Southern Planter,* I (1842), 13–14.

70.   *Ibid.*

71.   Although a fair was scheduled by the state agricultural society to be held at Jackson on November 20 and 21, 1844, it did not take place, and none was held thereafter until the latter half of the 1850s. Jackson *Southern Reformer,* December 6, 1844.

72.   In 1844 a Jackson newspaper editor commented on the decline of state and county agricultural associations: "An agricultural society is something like a physician, it is never called in until the planter has lost all hopes of his land." Jackson *Southern Reformer,* December 6, 1844.

73.   During the 1845–60 period, agricultural periodicals such as the *Southern Cultivator* and the *American Cotton Planter* received letters and articles written by Mississippians in constantly increasing numbers, and the same trend was discernible in the political press of the state.

74.   In 1845 Solon Robinson visited the plantations of a number of Mississippi cotton planters who had formerly been active in agricultural societies in their respective areas, and he found that Martin W.

Philips, John T. Leigh, Joseph Dunbar, James W. Towne, William Eggleston, John Hebron, Gadi Gibson, and numerous others were practicing agricultural improvement in a fashion that he thought highly pleasing. Kellar, *Solon Robinson*, I, 450–495.

FOOTNOTES FOR CHAPTER FIVE

1. Raymond *Times*, May 7, 1841.
2. A good description of the agricultural revolution that had occurred recently in Central Mississippi was contained in a letter written to N. G. North in February, 1842, by Colin S. Tarpley, *Southwestern Farmer*, I (1842), 59.
3. Raymond *Times*, August 2, 1839.
4. *Ibid.*
5. *Mississippi Free Trader and Natchez Gazette*, May 5, 1842.
6. *Southwestern Farmer*, I (1842), 19.
7. Raymond *Times*, September 24, 1841; *Southwestern Farmer*, I (1842), 46.
8. *Mississippi Free Trader and Natchez Gazette*, May 5, 1842; *Southwestern Farmer*, I (1842), 10–11.
9. *Mississippi Free Trader and Natchez Gazette*, May 5, 1842.
10. *Southern Planter*, I (1842), 7–8.
11. For biographical information on Affleck see Claribel R. Barnett, "Thomas Affleck," *Dictionary of American Biography*, edited by Allen Johnson and Duman Malone, I (New York, 1928–37), 110–11; Avery O. Craven, "The Agricultural Reformers of the Ante-Bellum South," *American Historical Review*, XXX (1927–28), 302–14; Wendell H. Stephenson, "A Quarter Century of a Mississippi Plantation: Eli J. Capell of Pleasant Hill," *Mississippi Valley Historical Review*, XXIII (1936–37), 355–56.
12. Jackson *Southron*, October 21, 1841; *Southwestern Farmer*, I (1842), 10–11.
13. *Mississippi Free Trader and Natchez Gazette*, December 2, 1841; Jackson *Southron*, December 9, 1841; *Southwestern Farmer*, I (1842), 10–11.
14. *Southern Planter*, I (1842), 10–11.
15. *Southwestern Farmer*, I (1842), 17, 44.
16. *Mississippi Free Trader and Natchez Gazette*, April 28, 1843.
17. In 1842, Philips wrote: "We want fine cattle, fine horses, and fine sheep; and we want to make large cotton crops." *Southwestern Farmer*, I (1842), 4.
18. *Southern Cultivator*, IV (1846), 56.
19. M. W. Philips to Samuel E. Bailey, December 6, 1841, *Southern Planter*, I (1842), 12–13.
20. *Southwestern Farmer*, I (1842), 19.

21. Riley, "Diary of a Mississippi Planter," Mississippi Historical Society *Publications*, X (1909), 373–74.

22. Colin S. Tarpley to N. G. North, January 30, 1840, Raymond *Times*, February 5, 1841.

23. Jackson *Southron*, November 11, 1841; *Southwestern Farmer*, I (1842), 22, 78.

24. *Southwestern Farmer*, I (1842), 17, 22, 44.

25. Kellar, *Solon Robinson*, II, 137.

26. Jackson *Southron*, November 3, 1841.

27. *Ibid.; Southern Pioneer and Carroll, Choctaw and Tallahatchie Counties Advertiser*, March 20, 1841.

28. Jackson *Southron*, December 9, 1841

29. *Mississippi Free Trader and Natchez Gazette*, May 12, 1842.

30. *Ibid.*, April 29, 1842.

31. *Ibid.*, May 5, 1842; *Southwestern Farmer*, I (1842), 10–11; Jackson *Southron*, December 23, 1841.

32. *Ibid.;* Kellar, *Solon Robinson*, II, 142–43.

33. A biographical sketch of Tarpley was published in *De Bow's Review*, XII (1852), 333–34.

34. Raymond *Times*, February 5, 1841; *Southwestern Farmer*, I (1842), 22.

35. Jackson *Southron*, May 25 and December 23, 1841

36. Riley, "Diary of a Mississippi Planter," Mississippi Historical Society *Publications*, X (1909), 312, 360.

37. *Ibid.*, 312–39.

38. *Ibid.*, p. 322.

39. Kellar, *Solon Robinson*, II, 128.

40. Jackson *Southron*, January 27, 1842.

41. *De Bow's Review*, XVII (1854), 627–28

42. U. S. Patent Office *Report: Agriculture* (1849–50), p. 161.

43. *De Bow's Review*, XVII (1854), 627–28.

44. *Ibid.*

45. Yazoo City *Yazoo Democrat*, September 24, 1845.

46. A more detailed account of the development of cotton and woolen mills in Mississippi during the depression can be found in: John H. Moore, "Mississippi's Ante-Bellum Textile Industry," *Journal of Mississippi History*, XVI (1954), 81–98.

47. Canton *Mississippi Creole*, March 26, 1842; Jackson *Southron*, March 3, 1842; *Mississippi Free Trader and Natchez Gazette*, May 5, 1842; U. S. Patent Office *Report* (1848), pp. 627–35; Raymond *Times*, February 5, 1841; *Southern Planter*, I (1842),12–13, 18–19; *Southwestern Farmer*, I (1842), 22, 87, 118, 121; Vicksburg *Sentinel*, January 24, March 10, and August 1, 1845.

48. U. S. Patent Office *Report: Agriculture* (1849–50), pp. 161–62.

49. *U. S. Census* (1840), p. 227; *U. S. Census* (1850), p. 456.

50. Jackson *Southron*, February 20, 1841.

51. Canton *Independent Democrat*, August 19, 1843.
52. U. S. Patent Office *Report* (1848), p. 509; U. S. Patent Office *Report: Agriculture* (1849–50), pp. 160–61.
53. *U. S. Census* (1840), p. 227; *U. S. Census* (1850), p. 456.

FOOTNOTES FOR CHAPTER SIX

1. *Southwestern Farmer*, I (1842), 59.
2. *Mississippi Free Trader and Natchez Gazette*, September 15, 1838.
3. *Ibid.*, May 30, 1839; *Spirit of Koscuisko*, May 18, 1839.
4. *Mississippi Free Trader and Natchez Gazette*, June 14, 1838.
5. *Ibid.*, May 16, 1839.
6. Raymond *Times*, May 7, 1841.
7. *Southwestern Farmer*, I (1842), 121.
8. Vicksburg *Sentinel*, December 14, 1850
9. These figures were compiled from the U. S. Censuses of 1840 and 1850.
10. *Southern Planter*, I (1842), 13; Vicksburg *Register*, February 20, 1838.
11. *Southwestern Farmer*, I (1842), 46
12. C. D. Bonney to Samuel E. Bailey, April 22, 1842, *Southern Planter*, I (1842), 13.
13. *Mississippi Free Trader and Natchez Gazette*, April 19, 1843.
14. Jackson *Mississippian*, March 11, 1849.
15. Riley, "Diary of a Mississippi Planter," Mississippi Historical Society *Publications*, X (1909), 403.
16. *Ibid.*
17. Martin W. Philips to Willie Gaylord, December 31, 1843, *Southern Cultivator*, III (1844), 62.
18. *Southern Cultivator*, IX (1851), 20; *ibid.*, XIV (1856), 162. Historical Society *Publications*, X (1909), 474 n.
19. Riley, "Diary of a Mississippi Planter," Mississippi Historical Society *Publications*, X (1909), 474 n.
20. W. P. Warfield to N. G. North, December 14, 1842, *Southwestern Farmer*, I (1842–43), 251.
21. U. S. Patent Office *Report: Agriculture* (1849–50), pp. 149–52.
22. Killona Plantation Journal, 1838, pp. 51–53, MS. in Mississippi Department of Archives and History.
23. Raymond *Times*, April 24, 1840.
24. Jackson *Southron*, February 20, 1841; *Southwestern Farmer*, I (1842), 1–2.
25. *Southern Cultivator*, I (1843), 202–203; *ibid.*, V (1847), 124–25.
26. Jackson *Southern Reformer*, September 6, 1845.

27. *Southern Cultivator*, V (1847), 124–25.
28. *Southwestern Farmer*, I (1842), 18.
29. *Ibid.*
30. *Mississippi Free Trader and Natchez Gazette*, June 14, 1843.
31. Jackson *Southron*, February 20, 1841.
32. U. S. Patent Office *Report: Agriculture* (1849–50), pp. 149–67.
33. Martin W. Philips to Thomas Eubanks, November 15, 1849,
U. S. Patent Office *Report: Agriculture* (1849–50), 149–52.
34. Wailes, Diary, March 19, 1853.
35. Vicksburg *Sentinel*, January 16, 1849.
36. Thomas Affleck to Thomas Eubanks, November 1849, U. S.
Patent Office *Report: Agriculture* (1849–50), pp. 153–54
37. *Ibid.*
38. *Ibid.*

FOOTNOTES FOR CHAPTER SEVEN

1. *U. S. Census* (1840), p. 227; *U. S. Census: Agriculture* (1850),
p. 456.
2. R. L. Allen, *The American Farm Book; or a Compend of
American Agriculture; Being a Practical Treatise on Soils, Manures,
Draining, Irrigation, Grasses, Grain, Roots, Fruits, Cotton, Tobacco,
Sugar Cane, Rice, and Every Staple Product of the United States
with the Best Methods of Planting, Cultivating, and Preparation for
Market* (New York, 1852), pp. 75, 163. Cited henceforth as Allen,
American Farm Book.
3. A. G. Ailsworth to N. G. North, June 14, 1842, *Southwestern
Farmer*, I (1842), 132.
4. John B. Stanley to Daniel Lee, August 1849, *Southern Cultivator*, VII (1849), 151.
5. Goodlee W. Buford to Daniel Lee, July 1, 1851, *Southern
Cultivator*, IX (1851), 116.
6. E. Jenkins to Daniel Lee, July, 1855, *Southern Cultivator*,
XIII (1855), 256.
7. *Southwestern Farmer*, I (1842), 109.
8. In 1846 Philips wrote to Colonel Alexander McDonald, of
Eufaula, Alabama, in regard to methods of soil conservation. His
method of using cowpeas for enriching the soil was "to flush deep,
cover shallow, push early, lay by early and sow peas at the rate of
one third of a bushel per acre, as early as I can lay by. The peas
cover the land, and assuredly have enriched my land. . . . I plant
thin land two years in corn [and peas] and one in cotton and feel
well assured that with peas I improve my land; and my neighbors
admit it. . . . My reliance has been on the cowpea. True, I use
cotton seed, stable, cow lot and hog pen manure; but what are all
these to 100 acres in cowpeas?" *Southern Cultivator*, IV, (1846),
78–79

9. M. W. Philips to N. G. North, 1842, *Southwestern Farmer,* I (1842), 116.

10. *Southern Cultivator,* V (1847), 90.

11. For a discussion of the cowpea controversy, consult the report of Thomas Affleck to the Commissioner of Patents, published in U. S. Patent Office *Report: Agriculture* (1849–50), pp. 158–59.

12. *Southern Cultivator,* XII (1854), 15–16.

13. *Southwestern Farmer,* I (1842), 205.

14. John B. Stanley to Daniel Lee, August, 1849, *Southern Cultivator,* VII (1849), 151.

15. *U. S. Census* (1840), p. 227; *U. S. Census: Agriculture* (1850), p. 456.

16. *Southern Planter,* I (1842), 15.

17. *Southwestern Farmer,* I (1842), 32.

18. Allen, *American Farm Book,* pp. 172–74; *Mississippi Free Trader,* April 19, 1843; *Southwestern Farmer,* I (1842), 121; U. S. Patent Office *Report: Agriculture* (1849–50), p. 159.

19. U. S. Patent Office *Report: Agriculture* (1849–50), pp. 149–52.

20. Allen, *American Farm Book,* pp. 155–56; *Southern Cultivator,* VII (1849), 107; *Southwestern Farmer,* I (1842), 113; *U. S. Census* (1840), p. 227; *U. S. Census: Agriculture* (1850), p. 456.

21. *Mississippi Free Trader and Natchez Gazette,* June 14, 1838.

22. Raymond *Times,* August 9, 1839.

23. *Ibid.,* July 19, 1839.

24. *Ibid.,* June 14, 1839.

25. *U. S. Census* (1840), p. 226–27.

26. Jackson *Southron,* November 3, 1841.

27. *Mississippi Free Trader and Natchez Gazette,* November 23, 1842.

28. Jackson *Southron,* June 1, 1842.

29. *Southwestern Farmer,* I (1842), 13.

30. *Mississippi Free Trader and Natchez Gazette,* June 14, 1843.

31. Moore, "Mississippi's Ante-bellum Textile Industry," *Journal of Mississippi History,* XVI (1954), 87–90.

32. This information was taken from a typescript copy of a unique farm diary written by Ferdinand L. Steel in the period between 1838 and 1845. The original manuscript is in the possession of Dr. Edward M. Steel, Jr., of Limestone College, Gaffney, South Carolina. Cited henceforth as Steel Diary.

33. Allen, *American Farm Book,* pp. 132–33; *Southern Cultivator,* IV (1846), 155.

34. Jackson *Southron,* March 17, 1842.

35. Martin W. Philips to Thomas Eubanks, November 15, 1849, U. S. Patent Office *Report: Agriculture* (1849–50), pp. 149–52; *Southwestern Farmer,* I (1842), 58; Wailes, *Report,* p. 187.

36. Jackson *Southern Reformer,* August 2, 1845; *Southwestern Farmer,* I (1842), 66, 90; Wailes, *Report,* p. 187.

37. U. S. Patent Office *Report: Agriculture* (1849–50), pp. 149–55; Wailes, *Report,* p. 188.

38. "Boston" to Daniel Lee, May, 1849, *Southern Cultivator,* VII (1849), 101.

39. *Southwestern Farmer,* I (1842), 154.

40. *Ibid.,* 58, 107; U. S. Patent Office *Report: Agriculture* (1849–50), 152–67.

41. *Southern Cultivator,* VII (1849), 85.

42. *Ibid.;* R. S. C. Foster, "History of Yellow Clover," *Southern Cultivator,* X (1852), 168.

43. *Ibid.,* 85.

44. *Ibid.*

45. Allen, *American Farm Book,* pp. 122–23.

46. Jackson *Southron,* December 9, 1841.

47. *Ibid.,* March 17, 1842; *Southwestern Farmer,* I (1842), 4, 92, 107, 153.

48. *Southwestern Farmer,* I (1842), 42.

49. U. S. Patent Office *Report: Agriculture* (1849–50), pp. 152–67.

50. Thomas Affleck, "Bermuda Grass," *Southern Cultivator,* III (1845), 91–92.

51. *Southern Cultivator,* VII (1849), 116.

52. Raymond *Times,* May 7, 1841.

53. Natchez *Courier,* December 26, 1849; New Orleans *Commercial Times,* January 21, 1846.

54. A detailed description of Affleck's home, farm and nursery was published in Thomas Affleck, *Affleck's Southern Rural Almanac and Garden Calendar for 1858* (New Orleans, 1858), p. 70.

55. "A Great Southern Orchard," *Southern Cultivator,* XVII (1859), 206–209.

56. *De Bow's Review,* X (1851), 602.

57. C. E. Hooker to the editors of the *Mississippian,* July 24, 1854, *Jackson Mississippian,* August 2, 1854. Refer also to *De Bow's Review,* XVII (1854), 626–30; Jackson *Mississippian,* December 9, 1857; *Southwestern Farmer,* I (1842), 44; Vicksburg *Whig,* November 22, 1853.

58. Jackson *Mississippian,* July 20, 1849; Natchez *Courier and Journal,* July 28, 1847; Port Gibson *Herald,* April 9, 1847; Vicksburg *Sentinel,* October 9, 1846, December 6, 1848, February 3, 1849.

59. Riley, "Diary of a Mississippi Planter," Mississippi Historical Society *Publications,* X (1909), 449; Jackson *Mississippian,* July 27, 1858; *Mississippi Planter and Mechanic,* XII (1857), 289; Raymond *Gazette,* September 2, 1857; *Southern Cultivator,* XVII (1859), 205.

60. Martin W. Philips to Edward Burke, November 6, 1848, U. S. Patent Office *Report* (1848), p. 508.

FOOTNOTES FOR CHAPTER EIGHT

1. The following extract from the *Alabama Planter* was published in the Columbus (Georgia) *Southern Sentinel*, April 11, 1850: "The vast importance of cotton as the leading staple of the Southern product has created an interest in the different varieties of the plant, and some considerable pains have been taken to produce new kinds, or introduce into notice and use those that have been discovered, and among these efforts some little humbug has no doubt crept in also. From what has occurred, we may safely consider the middle and southern portions of Alabama and the corresponding latitudes of this continent along the Atlantic and the Gulf, at a short distance from and parallel to them, as the true land of short staple cotton in its varieties of Green Seed, Mexican, Petit Gulf, Mastodon, and their cognate varieties. . . ."

2. Rush Nutt was conducting an experiment in hybridization at the time of his death in 1838. He had obtained seeds of a long staple black seed upland cotton in Egypt in 1834 and had cultivated it on his plantation near Rodney for several years. When the Egyptian cotton became acclimated, Nutt began to cross it with his Petit Gulf. His method of hybridizing was very primitive. He planted the different varieties to be cross-fertilized in alternate rows; the actual cross-pollination was left to the wind, insects and other natural causes. Nutt had planned to continue to mix the product of this process with fresh strains of Petit Gulf for several seasons, and after his death his son Haller Nutt carried out his instructions to the letter. Young Nutt finished the experiment in 1841, and gave samples of his "Egypto-Mexican" hybrid cotton to anyone that wished to test it. Unlike Richard Abbey or Henry W. Vick, he never sold seed from his breeding experiments; for he conducted them solely to satisfy his scientific curiosity. For details of the Nutt experiment in hybridization, refer to C. B. New to the editors, September 5, 1838, Rodney *Southern Telegraph*, September 12, 1838. See also Haller Nutt to Edmund Ruffin, April 25, 1841, Petersburg (Virginia) *Farmers' Register*, IX (1841), 312–14.

3. *Southern Cultivator*, XIII (1855), 216; Vicksburg *Whig*, February 27, 1856.

4. Jackson *Southron*, June 1, 1841; *Mississippi Free Trader and Natchez Gazette*, September 3, 1839.

5. A biographical sketch of Abbey was published in the *Biographical and Historical Memoirs of Mississippi* (Chicago, 1891), I, 278–80. A few of his papers are preserved in the Richard Abbey Collection, Mississippi Department of Archives and History.

6. Richard Abbey to James D. B. DeBow, July 15, 1846, *De Bow's Review*, II (1846), 132–42.

7. Richard Abbey to the president of the Agricultural and

Mechanic's Association of the State of Louisiana, December 20, 1845, New Orleans *Commercial Times,* January 10, 1846.

8.   Buckner and Stanton to Richard Abbey, November 22, 1845, and V. and L. G. Galloway to Richard Abbey, November 18, 1845, Jackson *Southron,* December 10, 1845.

9.   For a sample of Abbey's advertising, see *Southern Cultivator,* IV (1846), 141.

10.   Carrollton *Mississippi Democrat,* February 4, 1846.

11.   These testimonials were not always disinterested. The following advertisement was placed in the Fort Gibson *Herald* by a planter who had obtained experimental seed from Abbey:

"MASTODON COTTON SEED

I offer for sale a few hundred bushels of the Mastodon Cotton Seed, at four dollars per bushel. They may be had at Messrs. Summers, Vicksburg; McFarlan & Rundel, Grand Gulf; Stanton & Buckner, Natchez; and Buckner and Stanton, New Orleans, who have *not my written authority* to sell the seed, but plain consignments, seeking but fair competition. Neither is my signature on each sack, but my printed name. I do not deem it necessary to *Caution the public against* 'counterfeits,' for any one having eyes may see, and having hands may feel the difference who ever saw cotton seed. Neither do I deem it necessary to state that I did not 'name the cotton two years ago.' But this I do say that I warrant my seed as pure and genuine as any in the United States, and further that they are all saved perfectly dry; and from fully matured and not frost bitten bolls. I put up twenty-six pounds to the bushel.

Joseph Regan

Wyoming, Yazoo County, near Yazoo City, in the valley of the Yazoo River, January 10, 1846." Port Gibson *Herald,* January 29, 1846.

12.   New Orleans *Commercial Times,* January 3 and 10, 1846.

13.   Carrollton *Mississippi Democrat,* March 4, 1846; Port Gibson *Herald,* February 5, 1846; Vicksburg *Sentinel,* March 4, 1847.

14.   The details and results of a series of tests made with Mastodon cotton were published in a letter of M. W. Philips to John C. Jenkins, February 26, 1847, Vicksburg *Sentinel,* March 4, 1847.

15.   Martin W. Philips to Edward Burke, November 6, 1848, U. S. Patent Office *Report* (1848), pp. 505–509.

16.   Richard Abbey to the editors of the New Orleans *Tropic,* October 17, 1846, Jackson *Southron,* November 4, 1846.

17.   In 1855, A. W. Washburn, a farm implement inventor, cotton breeder and planter of Yazoo County, wrote in regard to Mastodon cotton: "I plant a small portion of my crop [in it] every year for late picking, and do not pick a lock of it until I have picked every

boll of the other kinds. I then, generally in January, gather it without pains, getting good weights; pass it through a very common gin, and sell it in New Orleans from twenty-five to fifty percent more than any other cotton." *Southern Cultivator*, XIII (1855), 340.

18. Washburn, and others, continued to improve Mastodon and sell the seeds during the decade of the 'fifties. In 1853, for example, he was getting two dollars a bushel for his "Improved Mastodon," which, he said, had "become acclimated, and improved in time of ripening and productiveness, without losing anything in length and fineness of staple." Yazoo City *Yazoo Democrat*, February 2, 1853.

19. Henry W. Vick to Martin W. Philips, March 15, 1847, Vicksburg *Sentinel*, July 7, 1848.

20. *Ibid.*

21. Vicksburg *Sentinel*, June 13, 1845.

22. *Ibid.*, July 7, 1847.

23. *Ibid.*

24. *Ibid.*

25. *Ibid.*

26. *Ibid.*, January 11, 1851.

27. Vick's Hundred Seed was the standard variety planted on "Nanechehaw Plantation" in Warren County as late as 1861. Charles Allen Plantation Book, April 6 and 13, 1861, in Mississippi Department of Archives and History.

28. Martin W. Philips, "The Different Varieties of Cotton Seed," *De Bow's Review*, XIX (1855), 224–25; J. V. Jones to Daniel Lee, January, 1850, *Southern Cultivator*, VIII (1850), 18–19; Martin W. Philips to Daniel Lee, March, 1850, *Southern Cultivator*, VIII (1850), 50.

29. *De Bow's Review*, XIX (1855), 224–25; *Southern Cultivator*, VII (1849), 11–12; *ibid.*, VIII (1850), 131–32; Vicksburg *Sentinel*, September 22, 1847, December 6, 1848; Vicksburg *Whig*, August 30, 1848.

30. *De Bow's Review*, XIX (1855), 224–25.

31. Martin W. Philips, "Varieties of Cotton Seed," *De Bow's Review*, X (1851), 567–68; A. W. Washburn to Daniel Lee, November, 1855, *Southern Cultivator*, XIII (1855), 340.

32. *De Bow's Review*, X (1851), 567–68.

33. The following advertisement appeared in the Vicksburg *Sentinel*, September 22, 1847:

"HOGAN COTTON SEED

We offer for sale a truly valuable article of COTTON SEED, price ten cents a seed. The Cotton from those seed has been examined by Judge Noland, and after examination he gave two hundred and fifty dollars for one fourth of a bushel. Doctor Noland examined the same, and purchased fifty dollars worth at ten cents for each seed. Various other gentlemen have examined it, all of who do not hesitate to say that it is the best article they have ever seen. The cotton is now growing at the residence of William Hogan in this county, and all persons wishing to purchase are requested to call and examine for themselves.

DOWNS & COMPANY."

34. G. D. Mitchell to W. S. Jones, April 30, 1852, *Southern Cultivator,* X (1852), 245–46.

35. William Hogan, "Hogan Cotton Seed," Vicksburg *Sentinel,* December 6, 1848. See also John Hebron, E. H. Bryan, and David Gibson, Jr., "Banana Cotton Seed, "Vicksburg *Whig,* August 30, 1848.

36. Aberdeen (Mississippi) *Independent,* February 1, 1851, and March 20, 1852.

37. Martin W. Philips to Daniel Lee, February 1, 1851, Aberdeen *Independent,* February 1, 1851.

38. Martin W. Philips to Daniel Lee, February 13, 1850, *Southern Cultivator,* VIII (1850), 41.

39. G. D. Mitchell to Daniel Lee, November 7, 1851, Aberdeen *Independent,* January 24, 1852. See also *ibid.,* March 20, 1852; Columbus (Georgia) *Sentinel,* January 3, 1850.

40. Aberdeen *Independent,* January 24, 1852; *Southern Cultivator,* VIII (1850), 92, 165; *ibid.,* X (1852), 245–46.

41. Martin W. Philips, "Hybrid Cotton," Yazoo City *Yazoo Democrat,* February 19, 1851.

42. *Southern Cultivator,* XI (1853), 79; *ibid.,* XIV (1856), 87 See also *De Bow's Review,* X (1851), 567–68.

FOOTNOTES FOR CHAPTER NINE

1. Grand Gulf *Advertiser,* February 8, 1839; *Mississippi Free Trader and Natchez Gazette,* August 3, 1837; Raymond *Times,* October 19, 1838.

2. F. Beaumont to S. H. B. Black, January 3, 1838, Vicksburg *Register,* January 9, 1838.

3. Panola *Lynx,* March 8, 1845.

4. *Southern Planter,* I (1842), 21–23.

5. A report of the Jackson planters' convention was published by the Vicksburg *Whig,* November 15, 1842.

6. *Southern Cultivator,* III (1845), 49–50.

7. New Orleans *Commercial Times,* June 20, 1846.

8. Vicksburg *Sentinel,* February 20, 1845

9. A statement of the number of bales of cotton delivered to Vicksburg by the Vicksburg and Jackson Railroad for the years 1847–54 was published in *De Bow's Review*, XVII (1854), 608.

10. The official returns of a census taken by the state in 1836 were published in the *Mississippi Free Trader and Natchez Gazette*, May 18, 1837.

11. *U. S. Census:* (1840), pp. 226–29.

12. *U. S. Census: Agriculture* (1850), pp. 456–60.

13. Good descriptions of the manner in which cotton was raised and processed for market during the depression period were published in *De Bow's Review*, II (1846), 132–42, and Allen, *American Farm Book*, pp. 201–203. Richard Abbey and M. W. Philips were the respective authors of these accounts.

14. For Philips's views on overseers and the overseer system of plantation management, see *Cotton Planter and Soil of the South*, n. s. I (1857), 348–49, and *Southern Cultivator*, XIV (1856), 339. His dealings with this class of employees are related in Riley, "Diary of a Mississippi Planter," Mississippi Historical Society *Publications*, X (1909), 311, 430, 436, 441–53, 456, 460.

15. For information on the wrought iron plows of the thirties, see M. W. Philips, "Plows," *Southern Cultivator*, V (1847), 90. For advertisements of the sale of such implements, see *Mississippi Free Trader and Natchez Gazette*, January 24 and February 2, 1837; Raymond *Times*, January 12, 1838; Vicksburg *Register*, January 4, 1838.

16. Martin W. Philips to John C. Jenkins, Jr., April 23, 1848, Vicksburg *Sentinel*, April 26, 1848. See also Jackson *Southron*, January 13, 1842.

17. Accounts of Philips's activities in testing plows were found in the following sources: *Mississippi Free Trader and Natchez Gazette*, December 2, 1841; *Southern Cultivator*, V (1847), 90, 153; Jackson *Southern*, March 10, 1842; U. S. Patent Office *Report* (1848), p. 509; Vicksburg *Sentinel*, April 26, 1848

18. "Bellows" [M. W. Philips] to the editor, February 24, 1842, Jackson *Southron*, March 10, 1842.

19. *Southern Cultivator*, IV (1846), 39.

20. *Ibid.*

21. In 1847 Philips remarked that "many castings brought to Mississippi are like pot metal. . . ." *Southern Cultivator*, V (1847), 90.

22. While passing through Yalobusha County in 1845 Solon Robinson saw in one field "20 one-horse, or one-mule plows, skinning the surface of this light, loose, fine, sandy soil, and sending it on a voyage to the Gulf of Mexico. . . ." Kellar, *Solon Robinson*, II, 449. In 1856 Edward R. Wells, a Yankee schoolmaster, saw much the same thing around Vicksburg. He recorded in his diary that "In some

lots twenty negroes with their roguish looking mules and *old-time* ploughs were scratching the surface of the earth. The sight is ludicrous to one accustomed to the subsoiling of Seneca County [New York]. It is the kind of cultivation that *wears out land. . . ."* Edward R. Wells, Diary, March 29, 1855. MS., Mississippi Department of Archives and History.

23. A detailed description of a harrow is given in *Southern Planter*, I (1842), 14.

24. Jackson *Southron*, March 10, 1842; *Mississippi Free Trader and Natchez Gazette*, August 2 and 4, 1838; *Southwestern Farmer*, I (1842), 42.

25. *Southwestern Farmer*, I (1842), 42.

26. *Southern Cultivator*, IX (1851), 36.

27. *Southwestern Farmer*, I (1842), 82.

28. The history of the "Mississippi Scraper" and descriptions of its various models were found in the following sources among others: *Southern Cultivator*, VII (1849), 51; *ibid.*, VIII (1850), 98; *Southern Planter*, I (1842), 20–21; *Southwestern Farmer*, I (1842), 49, 82; Vicksburg *Sentinel*, May 5, 1845.

29. V. N. T. Moon to N. G. North, November 25, 1842, *Southwestern Farmer*, I (1842), 116.

30. Information on gins was found in the following sources: Aberdeen *Independent*, April 9, 1853; Carrollton *Mississippi Democrat*, August 27, 1845; Jackson *Mississippian*, December 2, 1847, August 17, 1848, and February 3, 1854; Raymond *Times*, July 10, 1840; *Southern Cultivator*, XVIII (1860), 339, 343, 366–67; *Southern Planter*, I (1842), 22; Vicksburg *Sentinel*, April 13, 1850; Vicksburg *Whig*, August 27 and September 18, 1856; Wailes, *Report*, pp. 178–79.

31. For descriptions of different types of cotton presses used in Mississippi during the 1840–60 period, see Jackson *Southron*, April 22, 1846; *Southwestern Farmer*, I (1842), 77–78, 140; Vicksburg *Sentinel*, October 29, 1845; Wailes, *Report*, pp. 173–77.

32. Wailes, *Report*, p. 177.

33. Jackson *Mississippian*, April 20, 1849; Jackson *Southron*, December 23, 1841; *Southern Cultivator*, I (1843), 87–88; *ibid.*, V (1847), 104–105, 154–55.

34. "Caleb Quotem" [M. W. Philips] to the editor, April 19, 1849, Jackson *Mississippian*, April 20, 1849. See also *Southwestern Farmer*, I (1842), 149.

35. In a letter to Daniel Lee, written in November, 1847, Martin W. Philips reported the results of many years' experience with various types of fertilizers available in the cotton states. "We need not look to guano, lime, plaster, stable manures, peat, &c.," he wrote; "these

are too costly. We must either look to nature for the plan, or we must cultivate less per hand. . . . There is no doubt but what marl, lime, guano, and such articles would improve our lands much earlier than the old processes that our mother Nature has pursued for a few thousand years, but can we manure our lands at $10 to $25 per acre and sell corn at 30 cents per bushel, and cotton at 6 cents per pound? . . . where true economy is considered, we must often object to great advantage on account of cost." *Southern Cultivator,* VI (1848), 3–4.

36. In 1842, a Mississippi planter expressed himself on the value of barnyard manures on cotton plantations as follows: "Now there is nothing more common than to hear northern farmers inveigh against large farms—and it is for this reason among many, that their publications are inapplicable to our situation. They farm on a small scale and are forever dinning in our ears about the importance of barnyard and manure pens. But there is a vast difference between enriching a garden spot and a cotton plantation. Our want is the cheapest system of enriching and preserving large plantations—for it is outrageous humbuggery to talk of hauling manure over them. . . ." *Southern Planter,* I (1842), 12.

37. *Southern Cultivator,* II (1844), 45–46; *ibid.,* VI (1848), 117–18.

38. *Ibid.,* IV (1846), 134.

39. *Ibid.,* I (1843), 175–76.

40. In a letter to N. B. Cloud, written in September, 1856, N. T. Sorsby remarked on the subject of horizontal plowing: "I adopted the system when I first commenced planting in Hinds County, Miss., in 1844, where it was in common use. It had been adopted previously to that by my brother, John F. Sorsby, of whom I purchased a half interest in his land. We continued it without *guard drains,* or hillside ditches until 1851, when I moved to the fork of Greene County, Ala., and resumed it and added *guard drains.* . . ." *American Cotton Planter,* IV (1856), 347–48.

41. N. G. North, "Circling or Horizontalizing," *Southwestern Farmer,* I (1842), 17.

42. N. G. North, "Circling," *Southwestern Farmer,* I (1842), 73.

43. A detailed description of a common system of constructing guard ditches with given slopes appeared in a letter from M. W. Philips to William H. Jacobs, dated September 10, 1845, and published in the Port Gibson *Herald,* September 25, 1845. Accounts of slightly different methods in use in other Mississippi districts were published in the Woodville *Republican,* September 17, 1850, and *Southern Cultivator,* VIII (1850), 114.

44. *Southwestern Farmer,* I (1842), 73. See also *Southern Cultivator,* XIV (1856), 215.

240 AGRICULTURE IN ANTE-BELLUM MISSISSIPPI

FOOTNOTES FOR CHAPTER TEN

1. Donnell, *Chronological and Statistical History of Cotton*, pp. 370–501.

2. After carefully analyzing the manuscript census schedules for 1850 and 1860, Herbert Weaver concluded that Mississippi cotton growers—large and small, slaveowning and non-slaveowning—all shared in the general prosperity of this decade. Herbert Weaver, *Mississippi Farmers, 1850–1860* (Nashville, 1945), pp. 123–25.

3. *U. S. Census: Agriculture* (1850), p. 447; *U. S. Census: Agriculture* (1860), p. 270.

4. *Ibid.*

5. For the standard picture of the internal slave trade, refer to Frederic Bancroft, *Slave-Trading in the Old South* (Baltimore, 1931).

6. Phillips, *Life and Labor in the Old South*, p. 177.

7. *U. S. Census: Miscellaneous Statistics* (1860), p. 328.

8. *Ibid.*

9. *Hunt's Merchant Magazine and Commercial Review*, XLIII (1860), 632.

10. Moore, "Mississippi's Ante-Bellum Textile Industry," *Journal of Mississippi History*, XVI (1954), 81.

11. Although fire, weather and neglect have taken a terrible toll of Mississippi's ante-bellum homes, many of them are still in existence. Particularly fine specimens that have been altered but little can be seen and visited in the quiet old towns of northern Mississippi, especially in Oxford and Holly Springs. And a few plantation residences in varying states of repair are still standing in the surrounding countryside along almost forgotten roads.

12. Daniel Pratt, of Prattsville, Alabama, was the South's leading manufacturer of gins from 1840 to 1860.

13. Martin W. Philips, "A New Implement," *Southern Cultivator*, XIII (1855), 323.

14. *Ibid.;* Martin W. Philips to Noah B. Cloud, June 8, 1855, *American Cotton Planter*, III (1855), 244–45

15. G. D. Harmon, "Yost's Plow and Scraper," *Southern Cultivator*, XIV (1856), 274.

16. *Ibid.*

17. *Ibid.*, 364; Raymond *Hinds County Gazette*, April 23, 1856; *Southern Cultivator*, XVII (1859), 264.

18. Jackson *Mississippian*, April 16, 1856, and September 22, 1858; Vicksburg *Whig*, April 14 and October 14, 1856.

19. Jackson *Mississippian*, September 22, 1858.

20. Washburn's advertisements appeared regularly in the *American Cotton Planter and Soil of the South* and the *Southern Cultivator* between 1857 and 1860.

21. Martin W. Philips, "Scrapers," *Southern Cultivator*, XVII (1859), 199.

22. G. D. Harmon, "Winger's Scraper," *Southern Cultivator*, XVII (1859), 264.

23. Philips, "Scrapers," *Southern Cultivator*, XVII (1859), 199.

24. Harmon, "Winger's Scraper," *Southern Cultivator*, XVII (1859), 264. Refer also to *Mississippi Planter and Mechanic*, I (1857), 305.

25. Vicksburg *Whig*, March 10, 1858.

26. *Ibid.*

27. A broadside advertisement of the Stuart Double Plow and Double Scraper is included in the broadside collection of the Mississippi Department of Archives and History.

28. *Ibid.*

29. Oscar J. E. Stuart to John A. Quitman, August 29, 1857, John A. Quitman Papers, Mississippi Department of Archives and History.

30. Statements by a number of leading Mississippi planters included in Stuart's broadside relate the history of the implement.

31. *American Cotton Planter and Soil of the South*, n. s. III (1859), 270; Jackson *Mississippian*, July 27, 1858; *Southern Cultivator*, XVIII (1860), 259.

32. T. E. C. Brinley, "Dr. Philips's Plantation," *Southern Cultivator*, XVI (1858), 177; Martin W. Philips, "Southern made Implements," *Southern Cultivator*, XVII (1859), 39.

33. *American Cotton Planter and Soil of the South*, n. s. III (1859), 270; *Southern Cultivator*, XVIII (1860), 259.

34. *Ibid.*

35. *Ibid.*

36. J. W. Felt, "Test of Plows," *Southern Cultivator*, XVII (1859), 166–67

37. G. D. Harmon, "Plows," *Southern Cultivator*, XVI (1858), 208.

38. G. D. Harmon, "The Brinley and Other Plows," *Southern Cultivator*, XVIII (1860), 134.

39. Raymond *Hinds County Gazette*, September 23, 1857.

40. *Ibid.*

41. G. D. Harmon, "Hill Side Ditches," *Southern Cultivator*, XVI (1858), 214.

42. G. D. Harmon, "Grade and Horizontal Level," *Southern Cultivator* XVI (1858), 321.

43. Carlisle P. B. Martin to the editors, April 13, 1860, *Southern Cultivator*, XVIII (1860), 174.

44. Liebig's pioneering experiments in soil chemistry had attracted wide attention in the South during the 'fifties, and many of his letters had been reprinted in Southern agricultural journals. These letters

were collected and published in book form in 1859. John Blyth, ed., *Letters on Modern Agriculture, by Baron Von Liebig, with an Addenda by a Practical Agriculturist Embracing Valuable Suggestions, Adapted to the Wants of American Farmers* (New York, 1859).

45. R. D. Webb, "Science Applied to Agriculture—Vegetable Physiology," *Southern Cultivator*, IX (1851), 12.

46. *Ibid.*

47. Eli J. Capell to the editor, August, 1854, *Soil of the South*, IV (1854), 295. This letter was reprinted in *American Cotton Planter*, II (1854), 367.

48. *Ibid.*

49. For information on insect control methods current in Mississippi during the 'fifties, refer too: Aberdeen *Independent*, May 22, 1852; Duncan C. Hubbard to J. Holt, July 15, 1858, Okolona *Prairie News*, August 26, 1858; Yazoo City *Yazoo Democrat*, August 10, 1853; Wailes, *Report*, 147—49.

50. Hebron, "Boll-Worm, Sore Shin and Lice in Cotton," *Southern Cultivator*, XII (1854), 118–19.

51. A call for the organization of an agricultural society, dated July 6, 1852, that was published in the Yazoo City *Yazoo Democrat*, July 14, 1852, was typical of several appearing in the press at the time. Some of the older societies, like the Aberdeen Agricultural Society, had come to life and were laying plans for fairs to be held in the fall of 1853. Aberdeen *Independent*, June 11, 1853.

52. Martin W. Philips to the "Friends of Agricultural Improvement," September 10, 1853, Jackson *Mississippian*, October 11, 1853.

53. Martin W. Philips to Barksdale & Jones, November 19, 1853, Jackson *Mississippian*, November 25, 1853.

54. Jackson *Mississippian*, February 3 and 10, 1854.

55. Martin W. Philips to the "Planters of Mississippi," March 18, 1854, Jackson *Mississippian*, April 14, 1854.

56. For a discussion of the apathy displayed by Mississippians toward the "cause of agricultural improvement," refer to Thomas Affleck, "Southern Pomological Society," *American Cotton Planter and Soil of the South*, n. s. I (1857), 58–59.

57. Aberdeen *Independent*, November 20, 1852.

58. An account of the passage of this legislation was published in *De Bow's Review*, XXIII (1857), 639–40. The full text of the act was printed in the Jackson *Mississippian*, December 9, 1857.

59. Jackson *Mississippian*, December 9, 1857.

60. *Ibid.*

61. An account of the first meeting of the Bureau and an official text of its minutes appeared in the Jackson *Mississippian*, November 20, 1857.

62. J. J. Williams to the editors, Jackson *Misissippian,* May 26, 1858.

63. An article on the subject of Williams's tour of the state was published in the Jackson *Mississippian,* May 26, 1858.

64. Jackson *Mississippian,* June 9, 1858.

65. A synopsis of Secretary Williams's speech was published in the Kemper *Democrat* and reprinted in the Jackson *Mississippian,* June 29, 1858.

66. Jackson *Mississippian,* October 20 and November 17, 1858.

67. During October and November of 1858, local fairs were held at Paulding, Rodney, Canton, Raymond, Marion Springs (Lauderdale County) and other points in the state. For descriptions of these events, refer to: Jackson *Mississippian,* October 20 and November 3, 1858; Natchez *Free Trader,* November 1, 1858; Paulding *Eastern Clarion,* November 13 and December 4, 1858.

68. On the subject of local fairs, a Jackson editor remarked in 1859:

"Fairs in this state are of recent origin; they have been started within the last two or three years. The success which has attended them has exceeded the most sanguine expectations, and now they may be ranked as one of the permanent as well as most useful institutions of the State. . . . It must be remembered that the Fair is not only devoted to the exhibition of fine stock, of mechanical inventions, of agricultural productions, but it is looked upon as a convivial occasion. . . ." Jackson *Mississippian,* September 28, 1859.

For a similar expression of opinion, consult the Natchez *Free Trader,* November 5, 1859.

69. The best study of these cotton conventions to date is Weymouth T. Jordan, "Cotton Planters' Conventions in the Old South," *Journal of Southern History,* XIX (1953), 321–45.

70. Jackson *Mississippian,* August 3, 1859.

71. *Ibid.,* October 19, 1859, and July 17, 1860.

72. For the list of delegates from Mississippi and the attitude of the Bureau toward the problems to be considered at the meeting, consult Jackson *Mississippian,* September 6 and 28, 1859.

73. Jackson *Mississippian,* September 28, 1859.

74. *Ibid.,* April 17 and October 3, 1860.

75. Jackson *Daily News,* May 1, 1860.

76. *Ibid.,* April 24 and May 1, 1860.

77. Final arrangements for the state fair, which was held at Holly Springs in 1860, included no provision for competitions among military organizations. Jackson *Mississippian,* August 14, 1860.

Bibliography

I. MANUSCRIPTS

Allen, Charles, Plantation Book. Mississippi Department of Archives and History.

Allen, Henry, Papers. Mississippi Department of Archives and History.

Aventine Plantation Diary (1857-1860). Mississippi Department of Archives and History.

Brookdale Farm Plantation Record Book (1856). Mississippi Department of Archives and History.

Dunbar, William, Papers. Mississippi Department of Archives and History

Ellery, William R., Plantation Book (1855-1856). Mississippi Department of Archives and History.

Gordon, Stewart, Diary (1857). Mississippi Department of Archives and History.

Hunt, Abijah and David, Papers. Mississippi Department of Archives and History.

Jenkins, John C., Diary. Louisiana State University Library.

Killona Plantation Journal (1835-1844). Mississippi Department of Archives and History.

Quitman, John A., Papers. Mississippi Department of Archives and History.

Springfield Plantation Account Book. Mississippi Department of Archives and History.

Steel, Ferdinand L., Diary. In the possession of Edward M. Steel, Jr.

Wade, Walter, of Ross Wood, Plantation Diary (1845-1854). Mississippi Department of Archives and History.

245

Wailes, Benjamin L. C., Papers. Duke University Library.

Wailes, Benjamin L. C., Papers. Mississippi Department of Archives and History.

Wells, Edward R., Diary. Mississippi Department of Archives and History.

II. GOVERNMENT PUBLICATIONS

1. FEDERAL

American State Papers: Documents Legislative and Executive of Congress of the United States. 38 vols. Washington, D. C., 1832-60.

Bureau of American Ethnology, *Annual Reports, 1899–1900.* Washington, D. C.: Government Printing Office, 1899-1900.

Carter, Clarence E., editor, *The Territorial Papers of the United States.* 18 vols. Washington, D. C.: Government Printing Office, 1936–1954.

Census Bureau, *Aggregate Amount of Persons Within the United States in the Year 1810.* Washington, D. C., 1811.

——, *Abstract of the Return of the Fifth Census, Showing the Number of Free People, the Number of Slaves, the Federal or Representative Number; and the Aggregate of Each County of Each State of the United States.* Washington, D. C.: Printed by Duff Green, 1832.

——, *Compendium of the Enumerations of the Inhabitants and Statistics of the United States, as Obtained at the Department of State, from the Returns of the Sixth Census.* Washington, D. C.: Blair and Rives, 1841.

——, *The Seventh Census of the United States, 1850.* Washington, D. C.: Robert Armstrong, Public Printer, 1853.

——, *Statistical View of the United States: A Compendium of the Seventh Census.* Washington, D. C.: A. O. P. Nicholson, Public Printer, 1854.

——, *Population of the United States in 1860: Compiled from the Original Returns of the Eighth Census, Under the Direction of the Secretary of the Interior.* Washington, D. C.: Government Printing Office, 1864.

——, *Agriculture in the United States in 1860: Compiled from the Original Returns of the Eighth Census, Under the Direction of the Secretary of the Interior.* Washington, D. C.: Government Printing Office,1864.

——, *Statistics of the United States in 1860: Compiled from the Original Returns and Being the Final Exhibit of the Eighth Census.* Washington, D. C.: Government Printing Office, 1864

——, *Preliminary Report on the Eighth Census, 1860.* Washington, D. C.: Government Printing Office, 1862.

Department of Agriculture, *Yearbook of the United States Department of Agriculture, 1894–1954.* Washington, D. C.: Government Printing Office, 1895–1954.

——, *The Cotton Plant, Its History, Botany, Chemistry, Culture, Enemies, and Uses.* Washington, D. C.: Office of Experiment Stations, *Bulletin No. 33,* 1896.

Edwards, Everett E., *A Bibliography of the History of Agriculture in the United States.* Washington, D. C.: Department of Agriculture Miscellaneous Publication No. 84, 1930.

Patent Office, *Annual Reports of the Commissioner of Patents: Agriculture, 1849–1862.* Washington, D. C.: Government Printing Office, 1850–62.

2. STATE

Harper, Lewis, *Preliminary Report on the Geology and Agriculture of Mississippi.* Jackson, Miss.: E. Barksdale, State Printer, 1857.

Hilgard, Eugene W., *Report on the Geology and Agriculture of the State of Mississippi.* Jackson, Miss.: E. Barksdale, State Printer, 1860.

Report of the Professor of Geology and Agricultural Chemistry in the University of Mississippi. Printed by order of the Senate. Jackson, Miss., 1852.

Rowland, Dunbar, editor, *The Mississippi Territorial Archives. 1798–1803.* Nashville: Printed for the Mississippi Department of Archives and History, 1905.

Rowland, Eron, *Life, Letters and Papers of William Dunbar of Elgin, Morayshire, Scotland, and Natchez, Mississippi: Pioneer Scientist of the Southern United States.* Jackson, Miss.: Printed for the Mississippi Department of Archives and History, 1930.

Sargent's Code: A Collection of the Original Laws of the Mississippi Territory Enacted 1799–1800 by Governor Winthrop Sargent and the Territorial Judges. Jackson: Historical Records Survey, Works Progress Administration, 1939.

Toulmin, Harry, *The Statutes of the Mississippi Territory, Revised and Digested by the Authority of the General Assembly.* Natchez: Samuel Terrell, 1807.

Wailes, Benjamin, L. C., *Report on the Agriculture and Geology of Mississippi, Embracing a Sketch of the Social and Natural History of the State.* Philadelphia: E. Barksdale, State Printer, 1854.

III. CONTEMPORARY PERIODICALS AND MISCELLANEOUS AGRICULTURAL PUBLICATIONS

1. NEWSPAPERS

Aberdeen *Independent,* 1850–53.
Aberdeen *Whig and North Mississippi Advocate,* 1839.
Benton *Yazoo Banner,* 1838–44.
Canton *Independent Democrat,* 1842–44.
Canton *Mississippi Creole,* 1841–42.
Carrollton *Mississippi Democrat,* 1844–47.
Carrollton *Southern Pioneer and Carroll, Choctaw and Tallahatchie Counties Advertiser,* 1840–42
Columbus *Democrat,* 1836–52.
Columbus (Georgia) *Southern Sentinel,* 1850.
Grand Gulf *Advertiser,* 1835–39.
Grenada *Harry of the West,* 1844–46.
Grenada *Register,* 1842–43.
Greenwood *Reporter,* 1845.
Holly Springs *Gazette,* 1841–46.
Holly Springs *Guard,* 1842–46.

Holly Springs *Marshall County Republican and Free Trade Advocate,* 1838–39.
Holly Springs *Southern Banner,* 1839–41.
Jackson *Daily News,* 1860.
Jackson *Enquirer,* 1840.
Jackson *Mississippian,* 1835–59.
Jackson *Southern Reformer,* 1843–46.
Jackson *Southron,* 1840–50.
Macon *Intelligencer,* 1838–40.
Mississippi Free Trader and Natchez Gazette, 1835–60.
Natchez, 1830.
Natchez *Courier,* 1837–60.
Natchez *Mississippi Republican,* 1813–23.
New Orleans *Commercial Times,* 1846–49.
Okolona *Prairie News,* 1858.
Panola *Lynx,* 1845–46.
Paulding *Eastern Clarion,* 1858–60.
Paulding *True Democrat,* 1845–46
Port Gibson *Herald,* 1843–51.
Raymond *Hinds County Gazette,* 1849–60.
Raymond *Times,* 1837–42.
Ripley *Advertiser,* 1843–46.
Rodney *Southern Telegraph,* 1837–38.
Spirit of Kosciusko, 1839.
Vicksburg *Advocate and Register,* 1831–33.
Vicksburg *Constitutionalist,* 1844.
Vicksburg *Register,* 1838.
Vicksburg *Sentinel,* 1836–50.
Vicksburg *Whig,* 1839–60.
Woodville *Republican,* 1853.
Yazoo City *American Banner,* 1856.
Yazoo City *Yazoo Democrat,* 1844–54.
Yazoo City *Whig,* 1839–47.

2. AGRICULTURAL AND COMMERCIAL PERIODICALS

Augusta (Georgia) *Southern Cultivator,* 1843–60.
Albany (New York) *Cultivator,* 1834–60.
Baltimore *American Farmer,* 1839–60.

Baltimore *Niles Weekly Register*, 1811–49.

Columbus (Georgia) *Soil of the South*, 1851–56.

Jackson (Mississippi) *Planter and Mechanic*, 1857.

Montgomery (Alabama) *American Cotton Planter*, 1853–57.

Montgomery (Alabama) *American Cotton Planter and Soil of the South*, 1857–60.

Natchez *Southern Planter*, 1842.

New Orleans *De Bow's Review of the Southern and Western States*, 1846–60.

New York City *American Agriculturist*, 1842–60.

New York City *Hunt's Merchant's Magazine and Commercial Review*, 1839–60.

Petersburg (Virginia) *Farmers' Register*, 1833–42.

Raymond (Mississippi) *South-Western Farmer*, 1842–43

3.   AGRICULTURAL PUBLICATIONS

Affleck's Southern Rural Almanac and Plantation and Garden Calendar, by Thomas Affleck, Washington, Adams County, Mississippi. Mobile: M. Boullemet, 1851-58.

Allen, R. L., *The American Farm Book; or Compend of American Agriculture; Being a Practical Treatise on Soils, Manures, Draining, Irrigation, Grasses, Grain, Roots, Fruits, Cotton, Tobacco, Sugar Cane, Rice and Every Staple Product of the United States.* New York: C. M. Saxton, 121 Fulton Street, 1857.

Turner, J. A., editor, *The Cotton Planter's Manual: Being a Compilation of Facts from the Best Authorities on the Culture of Cotton; its Natural History, Chemical Analysis, Trade and Consumption; and Embracing a History of the Cotton Gin.* New York: C. M. Saxton and Company, 1857.

Wailes, Benjamin L. C., *Address Delivered in the College Chapel before the Agricultural, Horticultural, and Botanical Society of Jefferson College on the 24th of April, 1841.* Natchez: Privately printed, 1841.

———, *Address Delivered at Washington, Miss., before the Agricultural, Horticultural, and Botanical Society of Jefferson College on the 29th of April, 1842.* Natchez: Privately Printed, 1842.

IV. PUBLISHED MEMOIRS, DIARIES, JOURNALS,
TRAVEL ACCOUNTS AND CORRESPONDENCE

Baily, Francis, *Journal of a Tour in Unsettled Parts of North America in 1796 & 1797.* London: Baily Brothers, 1856.

Baird, R., *View of the Valley of the Mississippi, or the Emigrant's and Traveller's Guide to the West; Containing a General Description of that Entire Country; and Also Notices of the Soil, Productions, Rivers, and Other Channels of Intercourse and Trade; and Likewise of the Cities and Towns, Progress of Education, &c., of Each State and Territory.* 2nd ed. Philadelphia: Published by H. S. Tanner, 1834.

Baldwin, Joseph G., *The Flush Times of Alabama and Mississippi: A Series of Sketches.* San Francisco: Sumner Whitney and Company, 1876.

Bartram, William, *Travels through North and South Carolina, Georgia, East and West Florida, etc.* Philadelphia: Printed by James and Johnson, 1791.

Bassett, John S., editor, *Correspondence of Andrew Jackson.* 7 vols. Washington: Carnegie Institute of Washington, 1926–1935.

———, *The Southern Plantation Overseer as Revealed in His Letters.* Northampton, Massachusetts: Printed for Smith College, 1925.

Blyth, John, editor, *Letters on Modern Agriculture, by Baron Von Liebig, with an Addenda by a Practical Agriculturist Embracing Valuable Suggestions, Adapted to the Wants of American Farmers.* New York: John Wiley, 1859.

Carman, Harry J., editor, *Jesse Buel: Agricultural Reformer.* New York: Columbia University Press, 1947.

Cauthen, Charles E. editor, *Family Letters of the Three Wade Hamptons, 1782–1901.* Columbia: University of South Carolina Press, 1953.

Claiborne, John F. H., "A Trip Through the Piney Woods," *Mississippi Historical Society Publications,* IX (1906), 487–538.

Creecy, James R., *Scenes in the South, and Other Miscellaneous Pieces.* Washington, D. C.: Thomas McGill, Printer, 1860.

Cuming, F., *Sketches of a Tour to the Western Country Through the States of Ohio and Kentucky: A Voyage Down the Ohio and Mississippi Rivers; and a Trip through the Mississippi Territory, and Part of West Florida, Commenced at Philadelphia in the Winter of 1807, and Concluded in 1809.* Pittsburg: Cramer, Spear & Eichbaum, 1810.

Davis, Edwin A., editor, *Plantation Life in the Florida Parishes of Louisiana, 1836–1846, as Reflected in the Diary of Bennet H. Barrow.* New York: Columbia University Press, 1943.

Des Champs, Margaret B., editor, "Some Mississippi Letters to Robert Fraser," *Journal of Mississippi History,* XV (1953), 181–89.

Flint, Timothy, *Recollections of the Last Ten Years.* Boston: Cummings, Hilliard, and Company, 1826.

Fulkerson, Horace S., *Random Recollections of Early Days in Mississippi.* Vicksburg: Vicksburg Printing and Publishing Company, 1885.

Hall, Basil, *Travels in North America in the Years 1827 and 1828.* Philadelphia: Lee & Carey, 1829.

Hall, James, *A Brief History of the Mississippi Territory.* Salisbury, N. C.: Francis Coupee, 1801. Republished in Mississippi Historical Society *Publications,* IX (1906), 539–75.

Hogan, William R., and Edwin A. Davis, editors, *William Johnson's Natchez: the Ante-bellum Diary of a Free Negro.* Baton Rouge: Louisiana State University Press, 1951.

Hutchins, Thomas, *An Historical and Topographical Description of Louisiana and West Florida.* Philadelphia: published for the author, 1794.

Ideson, Julia, and Sanford W. Higginbotham, editors, "A Trading Trip to Natchez and New Orleans, 1822: Diary of Thomas S. Teas," *Journal of Southern History,* VII (1941), 378–99.

[Ingraham, Joseph H.], *The South-West by a Yankee.* 2 vols. New York: Harper and Brothers, 1835.

Keller, Herbert A., editor, *Solon Robinson: Pioneer and Agriculturist.* 2 vols. Indianapolis: Indiana Historical Bureau, 1936.

Ker, Henry, *Travels through the Western Interior of the United*

States from the Year 1808 up to the Year 1816. Elizabeth-town, N. J.: published by the author, 1816.

Lincecum, Gideon, "The Autobiography of Gideon Lincecum," *Mississippi Historical Society Publications,* VIII (1913), 443–519.

Mackenzie, E., *An Historical, Topographical, and Descriptive View of the United States of America and of Lower Canada.* 2nd. ed. Newcastle upon Tyne: Mackenzie and Dent, 1819.

Moore, John H., editor, "Two Documents Relating to Plantation Overseers of the Vicksburg Region, 1831–32," *Journal of Mississippi History,* XVI (1954), 31–36.

Olmsted, Frederick L., *A Journey in the Back Country.* New York: Mason Brothers, 1860.

Orr, J. A. "A Trip from Houston to Jackson, Miss., in 1845," *Mississippi Historical Society Publications,* IX (1906), 173–78.

Phillips, Ulrich B., editor, *Plantation and Frontier Documents, 1649–1863: Illustrative of Industrial History in the Colonial and Ante-Bellum South* (vols. I and II of the *Documentary History of American Industrial Society,* edited by John R. Commons). Cleveland, Ohio: A. H. Clark Company, 1909.

Riley, Franklin L., editor, "Diary of a Mississippi Planter," *Mississippi Historical Society Publications,* X (1909), 305–481.

Schultz, Christian, *Travels on an Inland Voyage through the States of New York, Pennsylvania, Virginia, Ohio, Kentucky and Tennessee, and through the Territories of Indiana, Louisiana, Mississippi and New Orleans, Performed in the Years 1807 and 1808.* 2 vols. New York: Isaac Riley, 1810.

Singleton, Arthur, *Letters from the South and West.* Boston: Richardson and Lord, 1824.

Smedes, Susan Dabney, *Memorials of a Southern Planter.* Baltimore: Cushing & Bailey, 1888.

Thwaits, Reuben Gold, editor, *Early Western Travels, 1748-1846: A Series of Annotated Reprints of Some of the Best and Rarest Contemporary Volumes of Travel, Descriptions of the Aborigines and Social and Economic Conditions of the Middle and Far West, During the Period of Early American*

Settlement. 32 vols. Cleveland, Ohio: A. H. Clark Company, 1904–1907.

Wills, William H., "A Southern Traveler's Diary, 1840," Southern History Association *Publications,* VIII (1904), 23–39, 129–38.

Wilkins, Jesse M., "Early Times in Wayne County," Mississippi Historical Society *Publications,* VI (1902), 265–72.

V. GENERAL HISTORIES, ENCYCLOPEDIAS AND BIOGRAPHICAL DICTIONARIES

Bailey, Liberty H., *Cyclopedia of American Agriculture.* 4 vols. New York: Macmillan, 1907–1909.

Biographical and Historical Memoirs of Mississippi. 2 vols. Chicago: privately printed, 1891.

Claiborne, John F. H., *Mississippi as a Province, Territory and State, with Biographical Notices of Eminent Citizens.* Jackson: Power & Barksdale, Publishers and Printers, 1880.

Craven, Avery O., *The Growth of Southern Nationalism, 1848–1861.* Baton Rouge: Louisiana State University Press, 1953.

Eaton, Clement, *A History of the Old South.* New York: The Macmillan Company, 1949.

Johnson, Allen, and Dumas Malone, editors, *Dictionary of American Biography.* 20 vols. and index. New York: Charles Scribner's Sons, 1928–1937.

Morison, Samuel E. and Henry S. Commager, *The Growth of the American Republic.* 2 vols. New York and London: Oxford University Press, 1942.

Rowland, Dunbar, *History of Mississippi: the Heart of the South.* Chicago-Jackson: S. J. Clarke Publishing Company, 1925.

Russel, R. R., *Economic Aspects of Southern Sectionalism, 1840–1861.* Urbana, Illinois: University of Illinois Press, 1924.

Simkins, Francis B., *The South, Old and New: A History 1820–1947.* New York: A. A. Knopf, 1947.

Sydnor, Charles S., *The Development of Southern Sectionalism, 1819–1848.* Baton Rouge: Louisiana State University Press, 1948.

VI. SPECIAL MONOGRAPHS

Alvord, Clarence W., *The Mississippi Valley in British Politics: A Study of the Trade, Land Speculation, and Experiments in Imperialism Culminating in the American Revolution.* 2 vols. Cleveland, Ohio: The Arthur H. Clark Company, 1917.

Bancroft, Frederic, *Slave-Trading in the Old South.* Baltimore: Johns Hopkins University Press, 1931.

Craven, Avery O., *Edmund Ruffin, Southerner: A Study in Secession.* New York and London: D. Appleton and Company, 1932.

——, *Soil Exhaustion as a Factor in the Agricultural History of Virginia and Maryland, 1606–1860.* Urbana, Illinois: The University of Illinois Press, 1926.

Davis, Charles S., *The Cotton Kingdom in Alabama.* Montgomery: Alabama State Department of Archives and History, 1939.

Demaree, Albert L., *The American Agricultural Press, 1819-1860.* New York: Columbia University Press, 1941.

Donnell, E. J., *Chronological and Statistical History of Cotton.* New York: published by the author, 1872.

Gray, Lewis C., *History of Agriculture in the Southern United States to 1860.* Washington: Carnegie Institute of Washington, 1933.

Hammond, M. B., *The Cotton Industry: An Essay in American Economic History.* New York: The Macmillan Company, 1897.

Jordan, Weymouth T., *Hugh Davis and his Alabama Plantation.* University: University of Alabama Press, 1948.

Mirsky, Jeannette, and Allen Nevins, *The World of Eli Whitney.* New York: Macmillan, 1952.

Owsley, Frank, *Plain Folk of the Old South.* Baton Rouge: Louisiana State University Press, 1949.

Phillips, Ulrich B., *Life and Labor in the Old South.* Boston: Little, Brown and Company, 1948.

Quick, Herbert, and Edward Quick, *Mississippi Steamboatin': A History of Steamboating on the Mississippi and Its Tributaries.* New York: Henry Holt and Company, 1926.

Rainwater, Percy L., *Mississippi: Storm Center of Secession, 1856–1861*. Baton Rouge: O. Claitor, 1938.

Ranck, James B., *Albert Gallatin Brown: Radical Southern Nationalist*. New York: D. Appleton-Century Company, 1937.

Stephenson, Wendell Holmes, *Isaac Franklin: Slave Trader of the Old South with Plantation Records*. Baton Rouge: Louisiana State University Press, 1938.

Sydnor, Charles S., *A Gentleman of the Old Natchez Region: Benjamin L. C. Wailes*. Durham: Duke University Press, 1938.

——, *Slavery in Mississippi*. New York and London: D. Appleton-Century Company, 1933

Watkins, James L., *King Cotton: A Historical and Statistical Review, 1790 to 1908*. New York: James L. Watkins and Sons, 1908.

Watt, Sir George, *The Wild and Cultivated Cotton Plants of the World: A Revision of the Genus Gossypium Framed Primarily with the Object of Aiding Planters and Investigators Who May Contemplate the Systematic Improvement of the Cotton Staple*. London, New York, Bombay and Calcutta: Longmans, Green and Company, 1907.

Weaver, Herbert, *Mississippi Farmers, 1850–1860*. Nashville: Vanderbilt University Press, 1945.

Wender, Herbert, *Southern Commercial Conventions, 1837–1859*. Baltimore: Johns Hopkins University Press, 1930.

Whitaker, Arthur P., *The Mississippi Question, 1795–1802: A Study in Trade, Politics and Diplomacy*. New York and London: D. Appleton-Century Company, 1934.

——, *The Spanish-American Frontier, 1783–1795: The Westward Movement and the Spanish Retreat in the Mississippi Valley*. Boston and New York: Houghton Mifflin Company, 1927.

VII. ARTICLES IN PUBLICATIONS OF LEARNED SOCIETIES

Bonner, James C., "Advancing Trends in Southern Agriculture, 1840–1860," *Agricultural History*, XXII (1948), 248–59.

——, "Genesis of Agricultural Reform in the Cotton Belt," *Journal of Southern History*, IX (1943), 475–500.

——, "Plantation Architecture of the Lower South on the Eve of the Civil War," *Journal of Southern History*, XI (1945), 370–88.

——, "The Plantation Overseer and Southern Nationalism as Revealed in the Career of Garland D. Harmon," *Agricultural History*, XIX (1945), 1–10.

Buck, Paul H., "Poor Whites of the Ante-Bellum South," *American Historical Review*, XXXI (1925), 41–54.

Carrier, Lyman, "The United States Agricultural Society, 1852–1860," *Agricultural History*, XI (1937), 278–88.

Coulter, E. Merton, "Southern Agriculture and Southern Nationalism before the Civil War," *Agricultural History*, IV (1930), 77–91.

Craven, Avery O., "The Agricultural Reformers of the Ante-Bellum South," *American Historical Review*, XXXIII (1927–28), 302–14.

——, "John Taylor and Southern Agriculture," *Journal of Southern History*, IV (1938), 137–47.

Hesseltine, William B., "The Mississippi Career of Lyman C. Draper," *Journal of Mississippi History*, XV (1953), 165–80.

Jordan, Weymouth T., "Cotton Planters' Conventions in the Old South," *Journal of Southern History*, XIX (1953), 321–45.

Leavitt, Charles T., "Attempts to Improve Cattle Breeds in the United States, 1790–1860," *Agricultural History*, VII (1933), 51–67.

McCall, A. G., "The Development of Soil Science," *Agricultural History*, V (1931), 43–56.

Moore, John H., "Mississippi's Ante-Bellum Textile Industry," *Journal of Mississippi History*, XVI (1954), 81–98.

Owsley, Frank L., and Harriet C., "The Economic Basis of Society in the Late Ante-Bellum South," *Journal of Southern History*, VI (1940), 41–54.

Seal, Albert G., "John Carmichael Jenkins: Scientific Planter of the Natchez District," *Journal of Mississippi History*, I (1939), 14–28.

Stephenson, Wendell H., "A Quarter Century of a Mississippi

Plantation: Eli J. Capell of Pleasant Hill," *Mississippi Valley Historical Review*, XXIII (1936–37), 355–74.

Swearingen, Mack, "Thirty Years of a Mississippi Plantation: Charles Whitmore of 'Montpelier,'" *Journal of Southern History*, I (1935) 198–211.

True, Rodney H., "The Early Development of Agricultural Societies in the United States," American Historical Association *Annual Report, 1920*, 295–306.

Whitaker, Arthur P., "The Spanish Contribution to American Agriculture," *Agricultural History*, III (1929), 1–14.

Index